Frederick Robert Augustus Glover

Harbours of Refuge

not

'Dangerous Decoys'

'Ship Traps'

nor

'Wrecking Pools'

Elibron Classics
www.elibron.com

Elibron Classics series.

ISBN 1-4021-4104-1 (paperback)
ISBN 1-4212-7522-8 (hardcover)

This Elibron Classics Replica Edition is an unabridged facsimile of the edition published in 1859 by Edward Stanford, London.

HARBOURS OF REFUGE;

NOT

"DANGEROUS DECOYS," "SHIP TRAPS,"

NOR

"WRECKING POOLS."

A REPRINT, IN PART, OF A PAMPHLET (DATED 1846), WITH SOME

ORIGINAL PAPERS ON MATTERS OF PRESENT INTEREST.

DEDICATED WITH GREAT RESPECT TO THE

LORDS COMMISSIONERS OF THE ADMIRALTY.

By FRED. R. A. GLOVER, M.A.

" The British Coast is the disgrace of the British Nation, and the Grave of the British Seamen."—Quoted by Calver.

" The Annual Loss of Property from Casualties on our Coast has been estimated at £1,500,000 sterling. * * * In one year alone, no fewer than 1549 persons perished from these causes alone."—Report of Harb. Ref. Comm., March 1859.

LONDON:

EDWARD STANFORD, CHARING CROSS, S.W.

M DCCC LIX.

NOTICE.

It is from no disrespect to a Gentleman, or to his useful and, as I think, meritorious labours, floating-harbour-wise, that all mention of Captain Adderley Sleigh's Floating System of Breakwater was omitted by me.

It was not until long after that page was printed, that I was aware that Captain Sleigh was prosecuting the Plan be presented to the Public many years ago.

I believe that a great deal that he says in his Pamphlet is very excellent, and I am sure, in accordance with Admiral Bullock, that moorings that would withstand any strain, may be placed, so to speak, anywhere.

QUESTIONS AND DEFINITIONS.

1. What is a "Harbour of Refuge?"

That into which a ship can sail and come to an anchor safely and lie in safety, under all circumstances and conditions of wind and tide; and also quit it when it suits her.

2. When is a Harbour of Refuge—so designated—a "Dangerous decoy?"

Every barred Harbour-so-called, is so, over the bar of which the water is not deep enough for the draught of your ship.

3. What is a "Ship-trap?"

Any (so-called) Harbour with One Entrance: the inseparable accident of which is, that a ship, having once entered (say for refuge from a storm), is detained until the wind, though quite fair for her voyage, changes.

4. What is a "Wrecking-Pool?"

A Harbour so-called; but, in reality, a Ship-Trap, as above, in which the swell is so violent, that ships will not hold their anchors, and wreck within the Piers in the Basin.

Where are these 2, 3 or 4, to be found? Everywhere that the Admiralty authority has been exercised.

Where are the Artifical Harbours of Refuge? Nowhere.

Where may they be? Anywhere that an Island Breakwater can be placed.

And where can such be placed? Possibilities considered, EVERYWHERE.

INTRODUCTION.

Wrecks increase, but not Harbours. Money is continually spent in *trying* to make harbours, and the result is, invariable failure. Mr. Calver calls our devices in this sort, "Dangerous Decoys." I call them, at least have called them, for a long time, Ship-Traps. They are both, and that in the worst sense.

Yet there is great talk about harbour-making, as if people really made, or meant to make, *Harbours*. We have, or have had, Harbours of Refuge Commissions, and Tidal Harbours Commissions, and Committees on Harbours, and Royal Commissions, and Roving Commissions to carry out views of Committees of the House, and Discussions at Engineer Institutes, and what not?—all about Harbours. No end of talk; and the thing is as far off apparently, and the wrecking as hopelessly irremediable, as ever.

Fourteen years ago the thing was committed to a Grand Harbour of Refuge Commission; and they committed it hopelessly to an eternal Slough of Despond. They set about so determinedly to tackle with the evil, that they declared they would do the thing effectually, once for all, immediately and eternally; and they laid out plans to do, and modes of doing, vast and grand; "walls of masonry" and nothing else! And so, Millions are wasted instead of Thousands used, and Ages consumed to do the work of Years and to meet the wants of To-day; and Wrecks go on, and Lives are lost, and Property is destroyed, and nobody profits save here and there an engineer. How long is this to continue? Dover, Portland, Harwich,—no end of money spent, nothing *done.* Into the famous port and harbour of Dover,—Who has not heard of the harbour of refuge at Dover? it has cost a million and a-half*—not a Packet can show itself at Low Water, nor land its passengers at all, if the wind be direct on-shore! Surely this is a frightful scandal. Who is to blame?

Twelve years ago I printed a pamphlet,† caused by circumstances which will appear in the sequel; I therein pointed out the folly of such a miserable scheme of building as was then in contemplation for that Site; at once, grand and ridiculous: but I was laughed at; even worse, sneered at. It is now my turn; and I mean to expose this wretched case by means of that very pamphlet which was held to be the produce of a fool or something worse. The Public—Members of both Houses, &c.—may look now for the "fool or something worse" elsewhere. Be it mine to give them the opportunity of judging for themselves, —of enabling them to descend upon the real Simon Pure.

I said to these people, "You talk of building cemented masonry out in the great deep, on a soft bottom, and you *can't* do it at all. And if you do it, by mere obstinacy of will, you will do it to a wreck. It will certainly cost you £10,000,000, possibly £20,000,000. You want the work *immediately*—you will be 20 years in doing it." *(See the speech quoted p. 8.)*

Now, the Engineer's Quarterly Reports make the first of these assertions but too palpable—a result of attempting impossibilities! —and they afford almost a guarantee of the second. The rate of cost exceeds even my estimate, and the result of 12 years

---

* The Scouring Works at Dover are said to have cost upwards of a million, before the recent outlay exceeding already £400,000 commenced. Those works were undertaken against the concurrent opinion of all the nautical men—and have proved to be money utterly wasted. See Adm. Elliott's evidence, 1971 to 2002; Admiral Boxer's, 1667 to 1697; Lieutenant Worthington's, R.N., 1825 to 1851, passim, Report of Dover Harbour, 1836, quoted in extenso infra.

† Entitled "Harbours of Refuge,"—CAVENDO TUTUS—not a copy of which can now be laid hands on except at the Clubs, British Museum, &c. 500 copies were printed and distributed. It is reprinted here, in part, Paper No. 8.

work is as above stated. Yet they might, instead of this, have completed this work for even less than merely one year's interest of the money they will have to spend; and it would have been completed out and out, and in full use for these last 10 years, instead of occupying 150 years to build it; for, a fourteenth part of the mural area of the work has taken 12 years to accomplish. And these are the doings of those who professed that a work on that Site was above all things necessary, and that it was wanted immediately! Who are the fools now? * * * Those who promise to do impossibilities? No; they are something else—but the men who believed those that deceived them. Chapter 8, THE REPRINT, (see p. 92,) will throw some light on the subject, and show that *the* engineering authority is just exactly where these wise men said it was not, and *is not* where the Public have hitherto thought, and the Authorities said, that it was. *Litera scripta manet.* Their promises are all "to the fore;" so are my assertions, which have been verified by the events. Let then the Public judge between us.

At a recent meeting of the Institution of Civil Engineers, "It was remarked," by one of the speakers, p. 13 (private paper), "that the subject of harbours of refuge, independently of the mechanical and engineering part of the question, was of vast importance, in a national point of view, whether as relating to the necessary protection of the commercial marine, or as connected with the naval defences of the kingdom. In order to rendering clear the points to be touched upon, it had been necessary to examine those formidable and not very lively documents, the Parliamentary Blue Books, and they confirmed the opinion, that all these great works were being executed without any efficient responsible supervision or controul; and that, not only the public, and the representatives of the nation, but also the Government itself, had been utterly in the dark as to proceedings relative to them. The time had now arrived when these matters should be brought before the bar of public opinion."

"The pier at Dover was stated to be now costing £415 per foot. At the present rate of progress, the projected harbour of refuge at Dover would scarcely be completed in less than 100 years, and at an outlay of £5,000,000. But the actual cost would not be represented by that amount; as, if the interest of that sum was spread over the hundred years consumed in the progress of the work, it would, with the principal, amount to £40,000,000."

"The renewed contract in 1854 was for 1000 feet at £415,000, or £415 per lineal foot,* and to be completed in 1864. It must be assumed that Parliament sanctioned that work, ordered it to be proceeded with, and voted the money on the report of the Commissioners: and yet, in 1858, on some of the Members of a Committee of the House expressing surprise at the slow progress of the work, and asking Mr. Walker this question—"Was it known, at the time it was decided to make the works at Dover, that it would take half a century to make them?" He replied, "I do not suppose it was. I do not think any idea was formed, at the time, as to the cost, or the mode in which it should be done,"†—and the eminent gentlemen should have added, "or, of the foundation which it was proposed should carry them."

So I affirmed and warned the public and the Government, (*ut infra*,) and the Duke of Wellington, in 1846; and for doing

* It is not to be inferred from this statement of price, that that will be the price to be paid when the bill is sent in; nor, that the subsequent portions of the work will be done for any such money. They have worked hitherto in comparatively shoal water; also not in an Island, making water carriage necessary. Now, they can avail themselves of land carriage. If now, it cost £415 per foot, and that irrespective of contingencies, I am very sure that Mr. Walker would not guarantee the completion of it at £1,000 per foot, liable to all contingencies. And it is well to bear this in mind, when the estimate of the speaker at the Engineers' Institute of £5,000,000 is calculated upon as the base of an ultimate accumulation of £40,000,000; for, if my amended calculation as to cost be realized, these figures will be doubled. A pretty bauble for our descendants to have to show for £80,000,000.

† Mr. Walker's *ipsissima verba* stands as the motto to Paper 7, p. 69

so, I was insulted and accused of insulting "eminent men;" and, now it comes out, that one of the Commissioners themselves, this same gentleman, recommended the doing of that, of which he had no idea "either of the cost of the work, or how to set about it." It is to be hoped, that this unblushing avowal of this Engineer Commissioner, who recommends works "without counting the cost," will make Members of the House of Commons cautious how they receive declarations of "no engineering difficulties" and promises to do impossibilities, and for given sums, when "no idea" exists as to the amount, nor how to do it, nor how to set about it, nor how long it will occupy. It therefore well became the speaker to conclude—"Hence it might be assumed, that Works were authorized, and the money of the Country voted away by the Government, without any idea being given of the time of construction, or of the cost of such works, nor even of their mode of construction. Now mark the result at Dovor: about £400,000 has already been expended, and yet it was, at times, nearly impracticable to effect the landing, at low water, of the passengers from the small steamer arriving from Calais, until the outward-bound steamer had left. This inadequate result, after such expenditure, was not creditable to the administrative skill of the Government of the country."

"In fine, it appeared that but little hopes could be entertained of more rational proceedings on the part of the Government, so long as Commissions were constituted as at present: because members of acknowledged independence were swamped by other members determined by foregone conclusions; and the schemes proposed by certain parties, were sure, by some 'hocus pocus,' to be, generally, recommended. It was evident that no person should be, placed upon such commissions who was pledged to any system whatever; nor should the execution of any work be confided to any member of a commission, by whom that work had been recommended. There were many modes of carrying out these views, and if a better and more equitable method was adopted, the best professional ability and the greatest executive skill, could be secured, to the manifest advantage of the country.

"This subject was one which well merited the attention of some independent member of the House of Commons, who might do good service to the country by pertinaciously attacking and exposing the present objectionable system of executing all government works."

Why the following Papers, 1 to 6, were written.

"RERUM COGNOSCERE CAUSAS."

SINCE the time that I lived upon the sea coast, and indeed from early life, I have taken a lively interest in matters nautical; but when it was proposed to make a harbour at Dovor, where I then resided, circumstances led me to conclude that the great difficulty would be much reduced, viz., that of building in 60-feet-deep water in the open sea, if the Government would adopt a Mode of Construction which I had been led to conceive.

Further consideration made me also conceive a Principle of Treatment for all River Mouths; by which, while I hoped Bar would be greatly abated, if not gotten rid of, I was certain that Access and Egress would be alway assured to the vessels wanting to get either into, or out of, harbour. Reading confirmed the view that I entertained,—the constant disasters at sea quickened my feelings,—and interest in the subjects made me study the Blue Books on Harbours, Light-Houses, Wrecks, &c. Hence, experience has enabled me to graduate to a good degree in the noble Science of Harbour Making, which I am not afraid to state, for I have out-lived the sneering; nor do I lack the power to maintain my opinion against all comers, as the contents of the following pages may suffice to shew. I accept as a great truth and encouragement what Mr. T. Stevenson is said to have stated: viz.—

"The construction of artificial places of refuge becomes a very important matter in a country where every winter's list of shipwreck and loss of life reminds us, how much Nature has left

Art to accomplish. The designing of harbours constitutes confessedly one of the most difficult branches of civil engineering:" "a fact," continues the ingenious gentleman who quotes his author, "which will be endorsed by the great majority of observers. The extraordinary differences of opinion," continues Mr. Calver, "which exist among professional men upon the point, evidently constitutes it a subject upon which little is known: for though, in other respects, we have made remarkable progress in all things of practical moment; Observation and Experience, in this one instance, seem to have signally failed to enlighten us; and we are still seeking after the truth with our ideas as vague, conflicting, and unsettled as ever."

"True for you;" certainly true: and the proof is, that after the vast quantity of money spent, and yards of long-winded "Reports" of "eminent engineers," and their "every confidence of success" in all their rash experiments in harbour treating,* there is *not one successful artificial Pier-Harbour in the Empire.* It is possible that Kingstown and Holyhead may both here be adduced to contravene this assertion. I have myself, seen, by the Charts, that the days of Kingstown are surely numbered; and at Holyhead, "in the Refuge Harbour, it was ascertained, at a late examination by the writer," Mr. Calver, R.N, "that, notwithstanding it is generally the receptacle of clear water, 15 to 18 inches of deposit have already formed in the deeper portions of it." In short, universal experience shews that it is "unsafe, and in disregard of all experience, to attempt to form a close Harbour within the limits of the heavily laden waters of England; where any close Harbour, must, in the end, prove only a stilling basin, a clarifying pool, and a mud-trap."† Yet these, with the addition of an

---

* There is a striking instance of the power of this "strong confidence" versus Common Sense hampered by gentlemanly good feeling, in the evidence of Captain Elliott, Dover Evidence, 2002.

† P. 43, "The Wave Screen." By Ed. Killwick Calver, Esq., R.N., Admiralty Surveyor, and author of "The Conservation and Improvement of Tidal Rivers.'

impassible Bar, constitute the staple of our harbours, our "ship decoys" of England.

Impressed with these convictions, and entertaining and expressing the IDEA that should change all this for the better, it was suggested to me that I should be examined by the Committee of the House that sat in May last. But failing in this, I addressed the following Papers to an hon. member, a gallant admiral, in the thought, that what I was not able to say to a Committee might be printed in an Appendix as was proposed. It was, however, too late; nor would they now have been printed, perhaps, had not they been submitted, as Document No. 2, for the consideration of the Royal Harbours of Refuge Commission; to which, by the great courtesy of the late First Lord of the Admiralty,* I had been referred. The Committee having received the instruction, that they were to afford me every facility for the illustration of my views on the subject, I accordingly submitted, as Document No. 1, the pamphlet already referred to; an abstract of which forms No. 8 of this Series: the next seven papers, were submitted as Document No. 2. To them were added—

A Plan of Treatment of the Port of TYNEMOUTH, as a work first proposed in 1846, and later in 1852.

A Plan for an Island Breakwater on the SMITHWICK SAND, in Burlington Bay.

A Plan for YARMOUTH ROADS; by which to convert them into a Grand Harbour, by raising the circumjacent Sands into Islands.

A Plan of an Island Breakwater for SOUTHWOLD in Suffolk.

A Plan, by which to save WEXFORD Harbour from being further ruined by mis-management.

A Plan for the creation of a Harbour in GALWAY ROADS.

A suggestion incidental for recovering the Navigation of the

---

* Sir John Pakington.

MERSEY, by reclaiming the ESTUARY of the DEE and MERSEY, and fixing its quicksands and re-accumulating the lost area.

And lastly, subsequently sent in, a Proposal for the Proper Treatment of the RIVER DANUBE, to save that river from a treatment for the recovery and improvement of its Navigation, which has failed wherever it has been tried—with the history of which transaction and the description of the proper treatment of it, and with a Postscript, which I find to be necessary, in consequence of the Recommendations in the Report of the Royal Harbours of the Royal Commission, dated March 1859, this publication will conclude.

This Fasciculum of Cases, with the proofs and matter incidental, would be the vehicle of communicating much valuable information to the public in a practical and interesting way, and the publication would form a collection of *Records.* I find it, however, too arduous to attempt such a work at present, and shall content myself with closing this brochure with the last case named, viz., the History of the Danubian Case, and the present condition of that vexed question.

In addition to the seven Papers of Document No. 2, I reprint as an Admiralty Episode, in proof of my case, a memorandum of the late Admiral Beechey, F.R.S., to the late Admiral Sir F. Beaufort, Hydrographer, "On the artificial harbours of the Irish Sea."

The Committee was supposed to occupy itself with discovering the best sites on which harbours of refuge should be constructed. Hence, as in the opinion of the writer, long ago expressed, Bridlington Bay was a most suitable place for such necessary accommodation, he was prepared with such evidence as the next paper gives, to maintain that opinion.

No. 1.*

Bridlington Bay and the Smithwick Sand. Reasons why a Breakwater should be erected upon that Sand.

Evidence from Mr. Edward Best, a great advocate for Filey Harbour.—" I believe that vessels going to the Baltic generally take their departure from Flamborough Head; and make it in returning home?" ." I believe they do." 2856.

" Therefore it is of great importance that there should be some safe harbour of refuge in the neighbourhood of Flamborough Head?" " Yes." 2857.

491. Do you like harbours placed in bights? Not at all.

493. Then, in considering the question of harbours of refuge, you would select the salient points of the kingdom as the most advantageous for harbours? Yes, for all purposes; for war, and as giving superior advantages to all vessels.

494 Both for purposes of defence and navigation? Yes; because if an enemy is seen off, a ship can be at sea in a moment.—*Captain Henderson's, R.N., Evidence*, p. 25, *Select Committee on Harbours.*

TIDAL Harbour Commissioners' Second Report, p. xii.:—

" BRIDLINGTON, like Whitby, has enjoyed a passing toll on coals for the last 150 years: like Whitby, too, it offers in return a harbour dry at low water; a north pier, built with public money at a very great cost, yet not even carried out to low water: no quay accommodation; a heavy range of swell along the pier, so that no moorings will hold the vessels; and a pier-end light, furnished with a single candle; yet *Bridlington Bay* is the only shelter in northerly gales on this coast, and in the

---

* It is to be understood that these next Seven Papers are, merely, the Papers as they were sent to Admiral Walcott; and the same as forwarded to the Commission. The mottoes have been added since.

continuance of such winds; 300 vessels may be seen lying here under shelter of the Smithwick Sand, which helps materially to break the sea, and offers admirable foundation for a breakwater, the construction of which would supply the great want of this part of the coast, by making a harbour of refuge of Bridlington Bay."

In the evidence given in the same Report, p. 389, No. 212, Mr. Harding, Treasurer of the Harbour of Bridlington Commissioners, "is of opinion," he having been 30 years a resident there, "that Bridlington Bay is the finest position for a harbour of refuge along the east coast of England, and is exactly the spot where such shelter would be most useful for the vast traffic that passes along the east coast of England. In the year 1811, he counted 450 vessels in the bay, and 1000 sailed past Flamborough Head in one day. If the vessels' ground tackling is good, they will ride a gale out in the bay, as the holding ground is good and the pull up-hill in on-shore winds. The Smithwick shoal, which extends across the mouth of the bay, with only 13 feet on it at low water, affords great shelter even in its present state; but if improved by a breakwater being placed upon it, the Bay would afford the finest shelter and the easiest of access on the east coast of England. It was a conviction of the value of the Bay that induced Parliament to renew the passing toll in 1837 for 21 years, in order to make an inner harbour, from which supplies to wind-bound ships could be sent off with certainty; and this, Bridlington, when its new south pier is finished, will be in a condition to do."

John Wallis, pilot and fisherman, 17 years at sea:—"The surge in the Harbour is very great; the anchorage in the Bay is good; no foul ground; 300 vessels left this week (25th October, 1845); in northerly gales, can anchor in the Bay in shelter when they cannot take the Humber. Vessels from the Baltic, bound to London and Hull, always make Flamborough Head in the fall of the year, and often anchor in the Bay. Berwick and

Hull traders also frequently anchor here in NNE winds. Vessels for Filey Bay also run for shelter here."

Benjamin Rawdon and John Lyons, pilots and fishermen—"confirm all that has been stated by Wallis respecting the Harbours and Bay: they consider Bridlington Bay *is* a harbour of refuge, and one of great value to the vast number of ships that pass along the coast; have seen 350 there at one time."

Mr. Thomas Brambles—"Lloyd's agent for ten years, is a native of Bridlington; has been 50 years at sea. Bridlington Bay is good holding ground, and the Smithwick Sand affords some shelter, yet 50 anchors and cables are lost in the Bay every year. A breakwater on the Smithwick is the great want: has long felt this, and has not ceased to press it on every occasion for the last 20 years: to say nothing of the lives that might have been saved, the property lost during the last half century for want of such shelter would have sufficed to build a breakwater: all his experience goes to shew, that a harbour of refuge, in Bridlington Bay, would be the greatest boon that could be offered to the vast number of shipping that trade along the east coast of England."—*Date of Evidence*, 1845.

Simon Goodrick, Esq., C.E., "Report of the proposed enlargement of Bridlington Harbour," Appendix B, p. 291, No. 213.

"Bridlington Bay, which is situate nearly in the middle of the eastern coast of the kingdom, is formed by the sudden projection of Flamborough Head on the north, and the less sudden projection of the coast towards Spurn Point on the south, and is thus protected by the land from winds and storms, from the north-east round by the north to the south-south-west, above half of the compass. The Smithwick Sand, about four miles in length, also forms a natural breakwater on the side of the Bay which appears open, and abates much of the violence of the storms upon it from the south to the east. The Bay has a good inlet and outlet at the north-east and the south, between the end of the Smithwick Sand and the land; and there is a good depth of water, safe anchorage, and ample space within it for large fleets; under these favourable circumstances, with the advantage of the light established on Flamborough Head, Bridlington Bay becomes a place of shelter and of great resort for ships northward or southward bound, meeting with contrary winds; so that it happens sometimes that 200 or 300 sail are lying in it, waiting for a wind." Mr. Goodrick's Report bears date September 1814.

Sir John Rennie. Select Committee on Shipwrecks, (July 1843,) p. 378, Admiral Dundas asks (5987):—"Do youk now any spot between the metropolis and the coal ports where a harbour of refuge could be made?"—"You might make one at Scarborough, and one at Burlington Bay." (5889.) "At what expense?"—"That I should be afraid to go into. At Scarborough there is an excellent depth of water, and the same in Burlington Bay." (5991.) "At no very extravagant price?"—"I should consider at a price not beyond its value." (5992.) "Five hundred thousand pounds?"—"That ought to go a great way."

[£500,000! I will undertake that two miles of work shall be erected for £200,000, and the harbour finished.]

Joseph Straker, Esq., North Shields, shipowner, dockowner, and shipbuilder. (June 1853.) (2439):—"Do you think it necessary to have any harbours of refuge on the east coast of England?"—"When a Committee of this House sat upon that subject some years ago, I was here as a witness from North Shields, and examined on that question; and I then gave it as my decided opinion, that the harbours of refuge that were then talked of were not desirable; but I beg to explain, that the harbours of refuge then proposed, were confined to a few localities: Bridlington Pier was one, Scarborough was another that was named, Whitby was another, and Redcar was another; and I then gave, as I give now, my evidence decidedly against those places, as harbours of refuge. I beg to correct that opinion now, so far as to say that my objection to harbours of refuge, as they were then proposed, was, because they were all situated in bights, or down upon the main shore. But I beg to say, that a harbour of refuge well to windward, would be a benefit." (2440.) "Where would you place that harbour of refuge?"—I think Bridlington Bay is one of the most desirable places for a harbour of refuge."

Mr. Robert Anderson, shipowner, resident at South Shields, Honorary Secretary to the Shipowners' Association. Select Committee on Shipwrecks, p. 124 (Question 1981):—"Have you turned your attention to harbours of refuge?"—"Not much: there is one place on the east coast of England that I have often thought would have made an excellent port of refuge, Burlington Bay, south of Flamborough Head." (2052.) "Would it be advantageous to the trade, if there were harbours of refuge on the east coast?"—"I think it would be very useful, if there was a harbour of refuge at Burlington Bay, and that it could be very easily made, by making a breakwater on the Smithwick Sand." (2053.) "At what expense?"—"I cannot say: but I think from the shoal water on which it could be made, it would be at much less expense than that at Plymouth. There is only nine feet of water on the Smithwick Sand. The Breakwater was made, I believe, in 12 fathoms." (2054.) "Have you been at Redcar?" —"I have." (2055.) "What is your opinion of Redcar?"— "It is situated in a Bight, and would be dangerous on that account: in my opinion, Burlington would be preferable to it."

Captain Washington. Select Committee on Shipwrecks, p. 101 (Question 1578):—"Are there not facilities for forming a harbour of refuge at Hartlepool?"—"Yes, Hartlepool is a very fair harbour. Hartlepool, at present, is one of the best harbours that we have there: and I think even under Flamborough Head, at Bridlington, it might be done. Flamborough Head affords a fine natural breakwater; but the place where a harbour of refuge seems to me to be most wanted, is Dover."

From the foregoing extracts from the evidence of the respective parties, it will be seen that Bridlington Bay receives,

c 2

concurrently, from those who are residents and have knowledge of the wants of navigation, and from those who live at a distance from it, and have no other interest in the place than as it commends itself to their judgment, the testimony, that it, of all other places on the coast, is best calculated, by means of a breakwater to be constructed on the Smithwick Sand, to aid the necessities of the storm-driven mariners on the East Coast.

The evidence of Mr. Straker and Mr. Anderson is conclusive against Redcar, as well as of Mr. Ballingall, given below:—"I think asylum harbours are most essentially necessary for the safety of merchant shipping, provided there be proper asylum harbours; [*i.e.*] that vessels could take at all times of tide, and in all states of the weather:" for this is expressing in general terms, what Mr. Anderson said expressly, in objecting to Redcar; that being proposed to be a one-entranced harbour, in an un-come-at-able Bight, and in accordance with the equally definite, as well as general, evidence against Redcar, and all Bight Harbours, given by Mr. Straker.

I have alluded to these objections to the proposed harbour of Redcar, because the present Hydrographer to the Admiralty, in his evidence in 1843, before the Select Committee on Shipwrecks, seemed then rather to incline to it. But that respected authority concurs equally with Mr. Straker, Mr. Anderson, Mr. Brambles, and Mr. Goodrick, that, at Bridlington, the construction of a harbour, such as would meet Mr. Ballingall's definition, namely, one "that vessels could take at all times of the tide, and in all states of the weather," and which Redcar certainly could not, "would be an advantage."

[I add the following evidence from the Hydrographer to the Admiralty, read by me since the foregoing was written:—(116.) "Then, coming further south, is there any point of the coast which is at present very much exposed to danger?"—"No; I think not. At Flamborough Head, in Bridlington Bay, a great many wrecks occur; it is a calling place for vessels in a northerly wind; I have known as many as 400 light colliers there. If a second harbour of refuge on the coast of England could be made, I should recommend one in Bridlington Bay, but not in preference to Tees Bay."

117. "If there were vessels on that part of the coast with a strong northerly wind, they could run for the Humber?"—"Yes, but it is a long distance to run for: vessels do not like to run so far to leeward of their direct course."]

Case of Filey Harbour versus Bridlington Bay.

Since the above was written, I have seen that Filey Bay has been started as a claimant to the honour of being the Port of Refuge to the bold mariners of England, when the fury of the storm compels Prudence to submit to the doctrine, that Discretion may be the better part of Valour.

It is wrong, but one did not really know where this Filey Bay was, and so, "The Sailing Directions" are consulted, and this is the modest account that its chronicler gives of it, under the head of "Flamborough Head." "Seven and a half miles from Flamborough Head is Filey Brigg, a bold, rocky promontory, that advances into the sea, forming a kind of hook; behind or to the southward of this hook, small coasters sometimes ride, sheltered from the N.W., but open to all other winds." And this is the place that is to be vamped up, at the cost of a computed £850,000, into a mercantile harbour; and, ultimately, at some fabulous increase of cost, into a Naval Rendezvous and Harbour of Refuge for the East Coast of England. As far as the place is concerned, it is nothing to me whether it be Filey Brigg or the Smith Wick Sand: the works at Filey as laid out, ought to be done for some of £250,000 instead of £850,000: this Dover and Portland rate extravagance, and "permanency" obtained by "works of masonry," will now, I suppose, after the Dover dose, be quite exploded. For, as Mr. Calver very judiciously says, p. 68.—"The difficulty and expense of constructing such harbours, have, in short, been the chief impediment to forming them; and some more economical plan is loudly called for:" something "based not only on intelligible principles, but, withal of so inexpensive a character, as would enable the number of refuge harbours round our sea-board to be multiplied twenty-fold." That is Common Sense and Humanity. As respects Filey Hook, however, it seems, that the respected Hydrographer of the Admiralty, who not only in July 1857, entirely disregarded the existence of Filey Bay, but recommended Burlington Bay as the site for a *second* harbour of refuge, has been contented, in the mean time, to adopt the recommendations which have been offered in favour of Filey Bay, and so, to stamp a certain amount of character on its pretensions.

Nevertheless, I take a view of the two cases on the *map*, (published by the Select Committee of Harbour of Refuge, 1857,) and I see this, first—that in one year, 1856, 3760 vessels *entered* Bridlington Bay; whence I infer, that a port or anchorage so immensely used, can hardly be said not to be worth protecting, especially when it can be easily done: and secondly, I see, that in a gale, easterly, and round to E.S.E., such a frightful sea would drive up into the proposed haven, that, when finished, it would be necessary to build another breakwater, E. from the shore, to keep vessels safe within it from being racked off from their moorings. I don't mean to say, that men might not shelter in this new port, even as it is declared they may; but I apprehend, that every one of them would be, sometimes, *right glad* to get out to sea again, yet be utterly unable to do anything but feel themselves going to irremediable ruin. Yet, such is the site concerning which, an admiral, an eminent engineer and commissioners, and captains of vessels, 99 out of 100, desire to see converted into the Grand Haven of East Anglia. This is asserted, but we haven't *seen* the opinions of the "99 out of the 100." We should be glad to have them canvassed with this view of the subject before them. Have they ever heard that even Kingstown Harbour is a Wrecking Pool? If not, they will know that striking fact before they finish this tract, from the evidence of a good many Steamer Captains: then, they may think differently.

In the mean time, the main argument relied on in favour of Filey, appears to be altogether founded on the comparison of Burlington Bay as it is, and Filey Hook—for it is *not* Filey Bay—as it is to be. It is said, that ships making for the south—all laden be it observed and able to hold the water—none of your "light ballasters" drifting to leeward—these *well-laden* ships—coming to the Head, and finding the wind south, and so, hindered *from going on their course*, toward London, &c., are compelled to drop back; and of these, on one ocassion, 300 were sheltered under Speeton Cliffs—half-way between Flamborough Head and Filey Bay; and very wise too, when they had no alternative but bad shelter under Speeton Cliffs, and none at all in Burlington Bay. But would they, one of them, have dropped back to Speeton had there been the object to gain, from a short struggle with the wind, of safety in a harbour or *sheltered* roadstead in Bridlington Bay?—the which, they might, every one of them, have been able to make, with a favouring tide?

All those vessels, which fall back into Speeton do so, not because they can't weather the Head, but because they know that if they did so, they would be worse off than if they did not. For, the reason that vessels, going southerly, don't now weather, or try to weather, Flamborough Head, is, just because there is no shelter for them if they succeed. The Smithwick Sand, with 9 to 16 feet water upon it at High Water Springs, in a furious Southerly Gale, is not a defence to be compared with the Smithwick Sand, with an Island breakwater upon it.

For a ship, under existing circumstances, to fetch into Bridlington Bay in a furious southerly gale, getting on for the top of a spring tide, would be downright madness, unless it was her captain's intention to cheat the underwriters by wrecking her. But, with the desired Work on the Smithwick Shoal, those 300 ships would have been added to the 3760 of 1856, if that were the year in which the reported event occurred. And if the recourse of ships be so surprising,—Mr. Coode has heard of 750 ships having brought up at one time under Speeton Cliffs, on account of the difficulties off Flamborough Head,—no doubt, plenty of Tugs would soon be found to tend the Anchorage of Smithwick, to help the heavy sailers round the point. But in the event of the *proper* wind setting into Filey Hook, when full of ships, no steamers would pull the ships out of *that* difficulty; they would be hardly able even to save themselves; and the rest of the shipping within, were they 100 or 1,000, would be all beached in one indiscriminate smash. The risk is too great; and the 1,500 masters and mates, and 1,500 seamen are too unheeding of probabilities, to make their petition of much value. And note, that during the time when such a frightful calamity as this would take place at Filey, not a ship would drag its anchor in Bridlington Bay, if *it* had its proper Work there. There is no occasion for us to be mystified with Petitions—look at the map! However, if there be a fashion set in for Filey Bay, by all means, take Filey Hook (for the *bay* is out of the question altogether); but don't let Bridlington remain without consideration. And indeed, if both be necessary, and Bridlington certainly is, let both be treated. Devote £250,000 to each, and have two harbours in some "*salient*" position; that will be a good deal better than having one, and that a Ship Trap, in the loop or hook of Filey Bay, at a cost of £850,000 *estimated*, and a Wrecking Pool withal: and, if not upon so grand a scale of magnificence in outlay as Kingstown, no doubt, quite as efficacious in its peculiar speciality of wrecking ships to perfection.

The arguments used in favour of Filey Hook, appear to me to be altogether to the advantage of Bridlington Bay.

Sir James Elphinstone asks—

490. "With respect to Filey Bay, you consider its principal advantage is, from its being a salient point?"—"Quite so."

491. "Do you like harbours of refuge placed in bights?"—"Not at all."

492. "Do you not consider them very frequently to be a trap?"—"As I mentioned, I have been with a very large ship in Mount's Bay, and, of course, I was always in terror if the wind should chop round, that I should not get out."

493. "Then in considering the question of harbours of refuge, you would select the salient points in the kingdom as the most advantageous for such harbours?"—"Yes, for every purpose, for war, and as giving superior advantages to all vessels."

494. "Both for purposes of defence and navigation?"—"Yes, because, if an enemy is seen off, a ship can be at sea in a moment. Captain Henderson, R.N.," (Evidence p. 25, Select Committee on Harbours.) I should be very glad to be told how Captain Henderson sailing frigates or sloops would get out of Filey Hook "in a moment," with the wind between S. and S.E.; and, as to his Mount's Bay terror, which he has so ably described, his Filey Hook Refuge would add to the excitement of Mount's Bay, that of having "terror" astern of him, from the moment he determined to be hooked at Filey. But, *de gustibus non est disputandum*. Captain Henderson, perhaps, *likes* Ship-Traps. It would seem that Sir James Elphinstone, by question 492, does not.

If a harbour *north* of the Head, is really so great a desideratum, and held to be entirely superior, to meet the difficulty at Flamborough Head, in position to Smithwick Sand Harbour, that already-*proved* Refuge of 3760 ships in one year, let it be constructed at Speeton; the use of which place is already proved, by 750 ships having been there seen at anchor; somewhat a different shewing from the half a dozen "small coasters," sheltering in Filey Hook. Yet this is the place for which £850,000 is modestly asked! As an old controversialist once asked of an opponent,

"Pray Sir, is that a joke or an argument?" So may one ask, "Is this a demand or a hoax?"

If then Bridlington Bay be the place for a breakwater, the next question is, how is it to be best constructed, and what will be the cost of such a work?

No. 2.

Island Breakwater Harbours, the only ones that are always accessible.

2282. "We cannot prevent wrecks." 2283. "But in diminishing the number of wrecks? No doubt, a harbour of refuge has a great tendency that way. No doubt about that, particularly such a one that ships can run in and out of, both ways."—*Admiral Evans, Admiralty Conservator of the Mersey, Evidence* 1858.

2481. Could a vessel always get out of Portrush if it were made a Harbour of Refuge? Yes; because there were two doors to it. There would be two entranees."—*Captain M'Keller, same Committee.*

"A detached pier or breakwater," by means of which "the harbours would be rendered safe and quiet in all winds," and "the means of egress and ingress would remain quite as free and easy as present, or even more so."

Mr. Cubitt's recommendation for the making Kingstown Harbour useable and safe.

In the first Select Committee on Shipwrecks, in 1836, the first witness called was a Mr. James Ballingall. He was examined on three different days, viz., the 1st, 5th, and 8th July; a man of such consideration that the Committee addressed to him 367 questions.

1. "What is your profession or occupation?"—"I am at present manager of a shipping company, and surveyor of shipping."

2. "Have you given particular attention to the extent of shipwrecks among the shipping of Great Britain?"—"Yes."

3. "For a considerable length of time, or only recently?"—"More particularly since the year 1830."

4. "Is it your impression that shipwrecks are on the whole more numerous at present than they were in former times?"—"Yes."

The last portion of this gentleman's evidence is as follows. After question 868 was answered, he volunteered,—"I was asked the last time I was here if there was any other suggestion I had to make to the Committee; it did not occur to me at the time, but

I think that asylum harbours are most essentially necessary for the safety of merchant shipping, provided there be *proper* asylum harbours," [*i.e.*] "that vessels could take at all times of tide and in all states of weather."..."In conclusion, I beg leave to state that under a better system I am most decidedly of opinion that more than three-fourths of the wrecks, as they are now taking place, might be with ease prevented; as I consider three-fourths are entirely attributable to the act of man, and not to the act of God."

To whatever causes Mr. Ballingall ascribes these effects, I beg to say that I most entirely and deliberately adopt his statement, as exhibiting the great fact of the case, with this modification, viz., that if three-fourths of these cases are attributable to the act of man, I should say that one-half are attributable to the neglect of man, arising from a perverse stupidity; and I point to the Wreck Chart, published yearly by the Board of Trade, to justify this my conviction and assertion; for, whatever amount of ships may founder at sea, owing to bad build, weakness of scantling, villainy from over-insuring, and the intentional wrecking of rotten ships, &c., the WRECK CHARTS *shew*—SHEW—that number ashore, or wrecked on, or at, or just off, the bars of the rivers, that, when they are counted up, they amount to nearly half of the whole loss: and there is no occasion to seek for deep, out-of-sight, causes for wreck, when the cause, in each such case, is as patent as the fact itself; namely, the inaccessible character of the port to which the mariner, in the exercise of his lawful business, had duly and effectively steered. These ships were lost, not from bad build, not from the villainy of false insurance, but because the Bar hindered them getting in to the haven were they would, and it was their business to, be. For, due observation of the Charts will shew, that these wrecks have all occurred in the endeavour of the ships to make ports where, *eternal shame to our engineers*, "vessels can" NOT "take, at all times of tide and in all states of the wind."

And what is the reason? because of the insanity or worse

which makes our "eminent engineers" continue to attempt impossibilities; driving eternally against Nature and shutting their eyes to her teaching: using the very means to create Bar, instead of availing themselves of her example and doctrine to reduce Bar: because, instead of learning and maintaining the doctrine of the double-entrance harbour, *i.e.*, an Island-defended harbour—accessible and egressible with every wind that blows—they persist in making Pier Harbours: that is, harbours with one entrance, which MUST have Bar before them. Harbours, so called, which, one while, wind off-shore, can't be entered; another while, wind on-shore, cannot be gotten out of: entrance into which, with a raging wind a-beam and corresponding sea, is a service of such danger as often causes a ship, half-in and half-out-of the Piers, receiving the full impact of the sea on her stern quarter, to be whirled round; when, losing her way and staggering astern, pressed down by her sails, she becomes a helpless mass and wrecks on the spot between the Piers; or, an equally common occurrence, mis-calculating the force of the moving waters, she steers not hard enough to the weather Pier, and thus, missing entrance, is drifted by the Port, and, before she has time to feel the effects of her helm, is wrecked on the lee of the Pier she couldn't make: wrecked, not by her own fault, but by that of these stupid unscientific devices. These occurrences are not less the common experience on the sandy shores, where open piers run out for hundreds of fathoms to sea, as at Ostend, Dunkirk, Calais, &c., than at the close solid stone piers, on our own coasts.*

---

* While these pages are going through the press, the inadequacy of such Shoal Ports has received a most painful illustration in the wreck at Calais, from a not very dissimilar cause, of the Mail Steam-Packet, "Frederick William." This is the sort of favour that we are about to confer on the St. George's mouth of the Danube by way of improving the Navigation. I trust, however, that my Lord John Russell will think it right, even yet, to put the veto on that hopeful scheme, even though the Great Powers have determined upon it. It will be no more mortification for the upholders of the present scheme to have to discontinue their works,

These *Pier* Harbours then, are, evidently, not according to Mr. Ballingall's type of what is wanted; but yet we attempt to make no other: we spend ten times the money that would make those others, but we won't make *them:* and here we are, after 14 years observation, and after all the pitiful attempts, "trials," that have been made, of *something,* to establish a System by which to make harbours to abate Wrecks, still in the midst of this miserable fact, viz., that—millions-worth of masonry and scores of infallible schemes to secure "water always over the Bar" notwithstanding—Shipwreck is increasing at the steady rate of 100 wrecks each year. They were 1200 the last recorded year—a few years ago they were 500: and, all this time, we have "Eminent Engineers" profuse in promises, never abashed though successful only in squandering vast sums of money, recommending the iteration and reiteration of the wretched and proved failures of 1000, and 100, and 10 years experience; the oldest, as well the latest, on record; abundance of "certainty" *assured;* none realized but the painful one of lavish Expenditure of Money and increased Waste of Life.

Here, for instance, is one "eminent engineer" who sets up the perfect harbour of Sunderland; and, (I speak under correction, if misinformed,) they get an act to squeeze out of the unfortunate shipping, "a passing toll" of some £500,000,* with which to construct this now, at last, *perfect* harbour of refuge,—*of Refuge;* and then one morning they look out to seaward, and find a glaring yellow streak right across the harbour, this harbour of refuge! They rub their eyes, and look again; but it won't disappear! Who pays for this broken promise?

---

than it was for the Galatz Commissioners to be ordered to undo what they had done; and there is clearly the same reason for objecting to one as there was for discountenancing the other. The Great Powers want the best Plan, not the exploded proved-failures of Europe.—See the concluding Paper in this Pamphlet, No. 10.

* This should have been £450,000 it seems.

So much for this harbour of *Refuge!* And this harbour of Safety—for "refuge" and "safety" are not at all one and the same thing that is, in the science of harbour-making in the school of the "eminent engineers,"—this harbour of *safety*, into which passing ships were to be enticed, what of it, after it had cost this £500,000?—what "safety" was there in *it*, when five ships wrecked and went to pieces *within it*, owing to the Swell in this perfect £500,000 harbour? Bitter reflection to aggravate the misery of the poor fellows, drowning alongside the steep quay walls of that cruel basin set up by quacks, that they had better have trusted to the tender mercies of the hurricane outside than have come in to more certain destruction in this curious contrivance for wrecking ships in safety—the *highly elaborated* work of eminent engineers! *

Nor is this all that this hopeful case has to present by way of instruction, nor its eminent engineer, whoever he is or was, has to answer for. The harbour, as has been said, was built, and it was *then* found to be unuseable by reason of the Swell. What was to be done? "Try something." "What shall we *try?*" "Let us try a Beaching Ground beyond the line of that north wall of the basin." So the £500,000 harbour is to be deprived of one whole side of its magnificent masonry, laid on splendid foundation; some two, three, four, or five more thousands of Pounds Sterling, are to be expended for removing this beautiful wall, and to excavate some acres of that very ground to a shelving beach, which, not long before, had been carefully filled in as backing to the now discarded wall; and so, were some 30 or 40 thousand Pounds Sterling utterly, scandalously, wasted. "Try that!" and after they had tried it, what happened? Why, what has been above declared; viz., that the five ships which had, by some mis-

---

* And there is some degree of resentment afloat against Mr. Calver, because he has designated these so-called Harbours, "Decoys."—See also Admiral Elliott's, Admiral Bullock's Lieutenant Worthington's, R.N., evidence already referred to.

chance for them, cleared over the Bar, had to rue their good luck of getting into the repose of the perfect harbour of Safety of Sunderland, with its latest improvements against wrecking.

One may be curious to know what became of that unwelcome Yellow Line, that stretched its ominous length across the harbour's mouth. "Oh, it was removed," as I was told. "How?" "Oh, it was dredged away." [How often, both before and since?]

"Notwithstanding," said another, a *friendly* informant to me, "it is steadily forming again; steadily growing." "Indeed; and what do the gentlemen say to that?" "Oh, it is never mentioned; they don't like to hear of it."

And yet, with this experience of failure at Sunderland, a similar sum of money, similarly levied,* for a similar purpose, was to be laid out at the Tyne, with just as much faith in the results promised as if all this had never been heard of; and another case of "try it again," to the tune of £500,000 is or was to come off.†

I think it is quite allowable to ask here, how long this tentative system of treating a vital want and vast subject is to be tolerated? I affirm that it is a melancholy exhibition of effrontery and quackery combined, that meets the reader as he "wades through the Blue Books" of the Tidal and other Harbours' Commissions. The whole history of the demonstrations of our "eminent engineers"—the honourable exceptions are not the favourites

* I speak under correction.

† It appears that this sum has been increased to £600,000, and now a further increase is needed, and an altered plan is provided. But why an *altered* Plan? If the Plan was well-matured in the first instance, why is it to be altered now? If the Plan of £500,000 is to be altered to £800,000, is it not another instance of the recklessness with which the Engineers act, who do not know, when Plans are proposed, either how they are to be set about, how long it will take to finish them, nor, even, what is proposed to be done? One man is hanged for looking over the hedge, while another walks off with the sheep with impunity. But then it is much to be "an Eminent Engineer." Withal, all that they *hope* to get is 16 feet. —Vetch, 934, 935. Mr. Calver *thinks* they *may* get 10, but nothing beyond that. A Harbour of Refuge, with ten feet over the Bar, to cost £800,000! and the whole thing is an *experiment*, avowedly so.—See Report of the Royal Harbour Commission, quoted on the Postscript.

of the authorities or of the public—is nothing more than a series of trials. "Try *this*." "This" does not answer. "Then try that." But "that" has already been tried, and has failed. "Then try something else." Thus, one man concludes that a north pier should be built: that not having produced the desired effect, with as little reason, in the next generation, a later eminence may say, try a south pier. Or, perhaps, one quack, bolder, or more impudent than the rest, says, "Try it again."*

Then an entrance is too narrow, and either the north or the south pier must be removed: or perhaps a new extension of both is asserted "with every confidence:" so a bold curvilinear work is to be run out into deep water to secure 18 feet soundings at low water springs. This is a bold experiment, but rash withal. The harbour is now useless by reason of Swell, and equally inaccessible with the former one, by reason of Bar. Then to check Swell, a fiddle-head, or kant, is to be put on this side, and failing here, it is to be tried there. And, that failing to give ease, a boom is to be tried,—and so, to do them justice indeed, every thing is tried but the right thing, an Island Pier—tried, in the vain attempt to do impossibilities, viz., to get rid of Bar at the entrance of a one-mouthed river harbour, and to have a close harbour, facing the sea, free from Swell.

The foregoing is about the experience, in part. of almost every River Port. As, in the words of a shrewd observer of the

---

* An eminent engineer has recommended that at Yarmouth a decayed north pier should be rebuilt; upon which Mr. Yetts, J.P. Yarmouth, wrote to Mr. Hume, one of the Commissioners who examined witnesses at Yarmouth, a month later:—"With respect, Sir, to our north pier, I have never been able to bring my mind to be convinced that because *three* north piers which had been created, and were allowed to fall into decay, on being found prejudicial to our harbour, that the construction of a FOURTH would necessarily be beneficial."

The *Italics* and CAPITALS are the worthy Justice's, not mine. I take the liberty to say that this gentleman's letters are well worth reprinting for universal reading by engineers: since, besides the very curious physical facts that they put on record, they may have the good effect of teaching eminent engineers how to observe, and that it is well sometimes to temper boldness with discretion.

doings at the Tyne, "It is dying of the doctor," so are most of these, dying of the doctor; or, what is the same thing, unfortunately without any figure of speech, the wretched mariners are being done for by the quacks; for the last stage of the cases that they undertake to handle is, mostly, worse than the first. Indeed, in one remarkable instance, an engineer of note had the courage and good feeling to say, "You have tried every thing and every effort has been equally unsuccessful; and things, you say, are worse now than they were. My advice is, that you let it alone." As if he had said: "The north pier and the south pier, and the lengthening, and the shortening, and the widening, and the narrowing have done nothing for you. What more can you do? Eat sand, and be content." This recommendation, however disagreeable to the applicants, had, at all events, the merit of honestly avowed inability to meet the case; and was a thousand times better than the "try this," of all the others, and that the doctor should not only aid in strangling the patient, but remorselessly send in a bill for the completion of the work.

But if all these efforts fail, as it must be admitted that they have failed, and *do* fail—for the long tail of wrecks at each Barred River's mouth, makes that but too evident—what is proposed to make things better? Simply, the construction of an Island Breakwater, before the mouth of every barred river; by which you will certainly realize the conditions so properly insisted upon by Mr. Ballingall, namely, a refuge "that vessels could take at all times of tide and in all states of weather."

But, perhaps, it will be said that this system of Island Breakwater that *you* propose is "tentative." If it be, I say it is better to try something that promises success, than trying that which is proved to be, and so is, certain failure. But it is *not* tentative. It was tried long ago, and has answered well. "The Creation of the World" is the date of structure, and the Record of Experience is through all ages, down to this very day—not read in books, but seen in the things themselves. What better Roadstead or more accessible than the Mother Bank behind the Isle of Wight?

"At Oban, the trade has wonderfully increased, because it has a bay, and well-sheltered with 100 acres of good anchorage," [Why ?] "because of being sheltered by the opposite Isle of Kerrara."—*Tidal Harbour Commissioners' Report*, xliv.

Loch Tarbert.—"At the western entrance of the loch is the Island of Trien, forming the harbour of Ardpatrick, in which the revenue cruizers will be found safely riding when the heaviest gales blow in the Sound of Jura."—*Ibid*, xlvi.

"Arran comprises the harbours of Lamlash, Brodeck, Corrie, and Loch Ranza. Lamlash, sheltered by the Holy Isle, is a spacious natural harhour, much frequented by the outward bound trade of the Clyde in strong westerly winds."—*Ibid*, xlviii.

"Ardrossan, at a projecting point, which forms the northernmost limit of the Bay of Ayr, is well sheltered by the Horse Island and outlying rocks, from all but south-easterly gales."

But it seems almost idle to quote instances to *prove* that islands are defences against the sea, and constitute Mother Bank, or Roadstead, always accessible, and that a ship can, under any wind, get behind them?

Why then is it that they are not universally adopted? Simply because as a Principle of Treatment they have not been thought of.

But now they *are* thought of, and have been for some years, by me, recommended. The next Paper will give an account of how my proposal to place an Island before the mouth of the Tyne was received by the engineering authority at that most important place.

[In the year 1844, I saw first the applicability of the System, in the case of Southwold, in Suffolk. Nature has there thrown up Hayle Sand, about ½ a mile long, just opposite to the mouth of the

River Blythe. On this Sand various cruel disasters have occurred. It is evident that if the Sand be made to grow into an Island, that the space behind it will become Sheltered Anchorage and the Sand Bank changed from an Enemy into a Friend. This doctrine is too deep for our Harbour Administrators ; so they continue to prefer the system that *must* wreck ships to the adoption of one that might save them. Light is, however, at length, breaking in upon their darkness. Captain Vetch now *thinks he sees or feels that* it is time for the darkness to make a change. Twenty years ago, he announced with some effect, that the Book of Nature was his study. In the year 1858, he *thinks* he may venture to say that an island is better than a pier harbour. This is at all events, a gain.

It was in the same year that I first proposed an Island Treatment of the Tyne difficulty, but it was not until 1852 that the occurrences of the next paper took place.]

No. 3.

Island Piers or Screens, and Jetty or Ordinary Pier Harbours, contrasted.

"The great art in constructing a harbour is to alter as little as possible, the existing state of things, and at the same time, to give the desired protection."—Sir John Rennie's Evidence, Select Committee on Shipwreck, 1843.

"It seemed to me that the late plans have been to a certain degree, only following up the old system of fighting against Nature, trying to overcome Nature which is not very easily done, and has failed in this instance."—Hon. Admiral Elliott's Evidence before the Committee for Dover.

"Much misapprehension exists as to the requirements of a Refuge Harbour—the repose of which is necessary in a dock for trading purposes—is not wanted, but only shelter from the tempest. It is necessary that a Refuge Harbour should possess ample space, and be so projected that vessels of the largest class would be enabled to enter and ride safely in it, under all circumstances of tide, wind, and weather, and that there should be also equal facility for quitting it. The matter of tranquillity is altogether secondary."—Calver, p. 54—5.

2257. "It is like being in Cork; they cannot get out again. The great object in a harbour, is, that when a ship is in, she may get out again in any wind. Now, there are two entrances to this, and whichever way the wind blows, they can get out if the ship carries sail," [or in.]—Admiral Evans's Evidence, 1858.

I HAVE insisted upon the desirableness of having AN ISLAND always in front of your river's mouth: when this is the case, Entrance and Egress is secure under all circumstances of wind, neither does Bar form in the Channels.

When, at the instance of His Grace the Duke of Northumberland, at the time First Lord of the Admiralty, I went to Newcastle,* and proposed to the Corporation of the Conservancy,

---

* I was not *sent* to Newcastle by the Duke. His Grace said, "You can try the Conservancy. They are very much interested in their River, and will be very glad, doubtless, to give attention to your plan."

the laying of an Island along and in front of the mouth of the Tyne, about half a mile from the shore; Mr. W. A. Brooks, the engineer of the Tyne navigation, said to me:—"Sir, it is evident that you are not a sailor. No sailor would think of such a thing as that!" Clearly, however, the Trustees of the harbours of Peterhead, N.B., are not of Mr. Brooks's opinion, for they, in the 15th paragraph of their Memorial, set forth that, "by making a cut between the two harbours to unite them"—in order to do which, they solicit aid—"vessels seeking shelter might have the singular advantage of entering the harbours or sailing from them, with the wind at any point of the compass."

Now, as Mr. W. A. Brooks, who inspires the decrees of the Tyne Conservancy, was believed by that confiding body, and I was, of course, disbelieved, the consequence was, the rejection of the proposition; and the corresponding consequence, that the usual number of ships have been yearly wrecked at the port ever since: notwithstanding that Parliament has enabled the Conservancy to mulct the shipping of the kingdom with a Passing Toll, to the amount of some £500,000,* which will have to be laid out according to the plans of this same eminent engineer, who rejects a plan on the very highest ground that can be given for its adoption; preferring to follow the evil example of Sunderland. It is hardly worth while to quote the opinion of the Peterhead trustees; but it reads curiously to a man who has heard of the rejection of a plan at the first river in England, as his Grace of Northumberland very properly declared the Tyne to be, trade wise;—rejected, precisely because it had those advantages which these gentlemen assert, and most correctly, to be of such "singular" value.

When Captain Bullock, R.N., Admiralty Surveyor, was asked (Select Committee on Shipwrecks, Questions 4353 to 4498):—

* I understood so, I think, from Mr. John Clayton, the very courteous Town Clerk of Newcastle, but I speak under correction.

4402. "What do you consider the qualities requisite for a complete harbour of refuge [on the part of the coast alluded to]?"—"A barrier from the prevailing winds, with a fair egress and entrance: I should call it a screen, rather than an harbour."

That is what is wanted. This is "The Island Breakwater:" Captain Bullock's "Screen."

4403. "Do you think one opening of 700 or 800 feet, in almost all weathers, supposing there was depth of water, would admit of a free ingress and egress?"—"I rather doubt 700, with large ships beating in or out." 4404. "Do you think it would with 1000 feet?"—"Hardly: it is very well for small vessels."

Here we see that a harbour with *one* entrance necessitates an opening of 1000 feet. On the other hand, in the case of Peterhead, a harbour which Mr. D. Stevenson, in conjunction with his father and brother, recommended to be adopted and improved for a Harbour of Refuge, by joining the two harbours, that so, the great advantage may be attained "of affording to shipping leaving the port, the choice of a north or south exit," and therefore, of course, the choice of a north or south entrance—the north entrance is 90 feet wide, and the southern 100. Now here it appears, even with *these* openings, "when there is any swell, it is hardly possible to hold a vessel at her moorings in the north harbour." "Vessels tear away from their fastenings." "Booms are wanted, or some other device to still the harbour." "The north harbour is uneasy to lie in."—p. 301, Appendix C. What, ships then could ride at anchor, in a close area having an opening to the ocean, rolling in all its power of surge through this opening of 1000 feet? [Consideration in favour of Filey Hook.]

"I once heard Captain Bullock say, myself, that he would at any time prefer to ride out a gale on-shore in Dover Bay, to venturing his ship into the basin between and behind the piers.*

---

* See a very remarkable Report or Memorandum, addressed by the late Adm. Beechey to the late Hydrographer, Sir F. Beaufort, on the Artificial Harbours of

"Why?" said I. "Because," was his answer, "I should be sure to wreck." I did not then understand what he could mean. *Now* I do; for I have since seen Sunderland, where five ships wrecked *in the basin* behind the piers, when they were well in (all lives lost also save one): they were wrecked by the Swell, and that, after the engineer had pulled down one whole side of the harbour, and converted the Quay that had been, into a Sloping Beach, in order that the Swell might exhaust itself against the sloping shore. Nevertheless, the work done, the work undone, the money spent and wasted, as much as would have paid for the creation of an Island Breakwater or Screen without, the five ships that *did* succeed in getting past the Bar, wrecked within the harbour of *safety* of Sunderland:—that model harbour of the eminent engineers, and which had been constructed at a cost of £500,000! Colonel Moody, R.E., and myself, were there not 21 days afterwards. [It seems this ought to have been £450,000.]

This sort of "raging of swell within" was also seen by my brother-in-law W. H. T. Hawley, Esq., of West Green House, Hampshire, (when Poor Law Commissioner of the district in the north,) to do like destruction in the harbour of Whitby; into which, an unlucky ship having entered, she rolled to that degree that the fire in the captain's cabin was hoven out of the grate; and the ship, taking fire in consequence, was soon a mass of

---

the Irish Sea. The whole Paper affords a wonderful development of unreasoning stolidity, counter-signed, of course, by the Admiralty, as backed by the judgment of all the authority of the high Sub-Officials. I think it well to reprint it as an Episode, "pour encourager les autres." It appears, in short, to be a powerful illustration of a General Principle, applicable to all the water-works to which the Admiralty "approved" has been set, which Mr. Farrel, C.E., has unwittingly pronounced, in announcing what he conceives is wanted at the Harbour of Courtown, in a letter to the Earl thereof.—"My lord, To render the opportunities that Nature afforded for Harbour purposes (at Courtown), available to the greatest extent, the works that have hitherto been erected, ought probably to be removed altogether." For, not only are Piers an abomination to Harbours as they are, but, when the Island Screen becomes to be adopted before them all, they will have to be removed as hindrances to the sailing vessels, endeavouring entrance.—See also the Dover Evidence, passim.

rolling flame from the deck to her vanes upwards, to the frightful peril of every thing within the harbour. She was herself burnt down to the water's edge. And these are the harbours for which our eminent engineers have so much admiration that they impose them *universally* upon the unprofessional mind of the Country and of Committees of the Honourable House! and *in the continued perpetration* of which, it may be, without exaggeration, said, that millions are yearly expended under sanction of Parliament.

Still, where these distressing wrecking-pools do exist, there is a remedy for the evil, if people will but open their eyes, look to evideuce, and "wade through Blue Books." In page 390, Appendix B., Tidal Harbour Commissioners' Second Report, Mr. Brambles, Lloyd's agent at Bridlington, says, "The range of swell along the pier is now so great that no ropes will hold the vessels: is of opinion, that nothing but *an island pier* will quiet the swell in the harbour: for this is the great evil; and fears that it will be still felt when the area is increased: no hawsers, and not even chains, will hold the vessels, so great is the range of the swell towards high water at spring tides."*

What is thus the recorded experience of Bridlington, of Sunderland, of Whitby, of Peterhead, &c., with respect to

---

* I have also found since these papers were written, that Captain Warren, R.N., poiuted out this as the remedy for Kingstown Harbour. How strange is it that Common-Sense and Seamanship is allowed thus ever and ever to bear witness in vain. I see also that Mr. Cubitt (p. 36, Appendix Tidal Harbour Commissioners' Report) recommended as Captain Warren did, "a detached Pier or Breakwater," by means of which "the Harbour would be rendered safe and quiet in all winds," and "the means of egress and ingress would remain just as free and easy as at present, or even more so." Captain Warren's opinion is as follows:—

"There are various opinions as to the plan that would most likely prevent the swell which sets into the harbour with an easterly wind. I consider that any thing carried across the set of the tide, would cause an eddy, and be likely to form a Bar at the entrance. The most effectual plan would be another breakwater, placed at a distance of about two cables length from the Pier Head, in the direction of the tides."—Captain Warren's, R.N., Report on Kingstown Harbour.

Swell, is that of *every other* pier harbour into which the ocean rolls when the wind is on-shore. What Mr. Brambles suggests as the remedy for Bridlington, is the same as Captain Bullock's Screen, and is a modified realization of the improved—perhaps by this time restored—harbour of Peterhead, with its beautiful natural great rock-island-breakwater, when the cut shall be made which is to make one of the two now separated harbours: an harbour or "harbours," as it is expressed, "having the singular advantage of being able to enter them and sail out of them with the wind at every point of the compass;" and that, also, with openings to the sea, *half or quarter of the width of entrance necessary when there is a harbour with one opening.* It is to be hoped that this "singularity" of advantage, on which the Peterhead men are so properly disposed to plume themselves, will henceforth cease to be a peculiarity of that place so happily favoured by nature.

"If the cut were made, all would seek shelter in the south harbour."—Alex. Souter's evidence. But this is not the main point here under consideration. When the in-rolling sea could go *through* the harbour, the swell and surge would, certainly, be greatly abated.

It is, however, mainly true, as Mr. Souter says, "the greatest evil at Peterhead is the want of a cut between the two harbours. From want of this, has known vessels detained days and sometimes weeks." Other people might be expected to see this and appreciate a double entrance as well as Mr. Souter, although "no sailor would think of such a thing as *that*." Yet Mr. Souter it seems has been 36 years at sea. He adds, "There *was* a passage for the sea formerly, at high water, between the two harbours, but the mound, called the Sand Ridge, was formed to connect the islands with the town." Not to compare small things with great, it is remarkable that the harbour of Tyre, the Queen of the Sea, was destroyed by a similar sand-bridge, caused to be constructed by Alexander the Great for a like purpose.

In the foregoing collection of data for the purpose of exhibiting the advantage of the Island Breakwater or Screen, over Piers in prolongation of the sides of the rivers, while I have used Peterhead as a case by which to illustrate my assertion, the neglect of the True Remedy for the Evil of Pier Harbours, is variously illustrated in the same Report of the Tidal Harbour Commissioners; where it appears that, while Bar and Swell are the common nuisances of all Pier Harbours, the presence of an Island always seems to secure good harbour and easy access.* I have already mentioned, especially, the case of Oban, of which the bay is sheltered by an island. But a very remarkable instance of the value and effect of an Island before a river's mouth, and the difference that resulted from changing the double entrance to a single one, is that of Wismar in Mecklenburg. This navigation, screened by a small island, and used by the Swedish admiral at all times, was taken by the Danes, who desired to destroy it. To this effect, they filled in old ships and matter between one end of the island and the main. Bar immediately formed before the other entrance, and the port became utterly destroyed in character.

I beg to call the attention of the Committee to this very remarkable case. It shews not only how true Nature is to herself, in that a harbour good for centuries became at once inaccessible, when the cause was removed which gave a double channel for the exit of the river to the Baltic; but—what is very much to be remarked upon, viz.,—the superiority of engineering science-physical in the Danish Engineers in the beginning of the last century, to that of all our "eminent engineers" 150 years later. Be it remembered, the Danes, *with science*, destroyed that island-defended entrance, the use of which, we, with all our high appreciation of ourselves, have not yet been able to comprehend."

---

* If Boddam be not "easy of access," that is owing to the rocky nature of its approaches. It is safe, when once in.—Tidal Harbours Report.

This case is worth investigating. It also affords us an opportunity to guage the staple of our River Conservancies. When I was about to appear before the Royal Harbour Commission in December last, I was requested, if possible, to obtain the proof of this case of Wismar.

What I know about this case is as follows:—In the year 1847, in a conversation which I had with the celebrated Professor Arndt, in Bonn, when pointing out to him the desirableness of treating all harbours island-wise, the learned gentleman answered,—" Oh, there's nothing new in *that*. The Danes knew all about *that* 150 years ago, when so and so (as above) took place." Being anxious later to ascertain the details of this operation, I applied, about two years since, to the learned Professor, who said,—" The father of Professor ——— was long Burgomaster there, and his brother is Town Councillor (Stadt-rath) there now; he will gladly tell you all the particulars."

On the recent requirement of the Commission, alluded to, I again applied to that learned person, who not only knew nothing at all about it, but ignored the Professor's facts, and denied that Wismar was a bad port at all. He, evidently, did not approve of his town and birth-place being placarded to all the world as the worst port in Europe; a feeling highly creditable in the patriotic, of however questionable propriety in the cosmopolitan, sense. Howbeit, the learned gentleman so stoutly negatived this last view of the specialties of Wismar, that had I not had my doctrine established upon something stronger than the evidence of the venerable nonagenarian, I should have felt myself, notwithstanding my entire confidence in my authority, rather in the position of having imposed fancies for facts on the attention of the Commission. But I *knew* that the Harbour was a bad one, from what follows :—

On the memorable occasion when I had to " incense " each of the Conservators of the Tyne with my " views " on the necessary treatment of their river's mouth, it was my fortune, good or bad, to have to work my way into a gentleman conservator's office, who lived, or was to be seen, in a house on the Quay at Newcastle. "May I ask your business, sir?"—"Oh yes; I am anxious to speak to you concerning the treatment of your river's mouth."—"What do you propose, sir?"—"Why," said I, "as in the treatment of the river at Wismar they did——"—"Wismar, sir! Wismar, sir! Bless my soul, sir! the worst port in Europe, sir! I have had enough of Wismar, sir, too much: I have had a ship there six weeks trying to get in; and once in, three months before she could get out! Make the Tyne like Wismar, sir! like the worst port in Europe, sir! never heard of such a thing!" The poor gentleman hereupon worked himself up into a state, such that it was hopeless to attempt to make him understand that I had quoted Wismar as a beacon to avoid, so I left him in the full persuasion that I was about to undertake to make the mouth of the Tyne a true copy of "the worst port in Europe." These facts I conclude, however arrived at, will be sufficient authentication of the venerable Arndt's statement, that the harbour into and out of which the Swedish admiral before the year 1715, with his hundred gun ship, could sail without let or hindrance, is now, what he declared it to have become, in consequence of the doings of the Danes with malice prepense, viz.;—utterly spoilt; the opinions of any interested persons to the contrary, notwithstanding.

No. 4.

Ballasting and Dredging: Material for the Island Breakwater: the Nuisance of, abated by this Means.

"For the improvement of Harbours there is no single thing that has been attended with such beneficial results as the free use of the steam dredge: in nine cases out of ten this is the grand cure for the general evil complained of, namely, want of depth. * * * * Nor would it be easy to point out how a greater benefit could be obtained at so small a cost."—Captain Washington's Report, p. lxii.

There is another somewhat prominent point from which to view with regret the inability of the Tyne Conservancy and their professional adviser, to see the applicability of the work proposed by me, for the Tyne, some years ago: because, besides the great advantages promised in the way of navigation, and also of stilling the anchorage up the river, it would have gone a great way, for a great many years, to abate the often destructive, always onerous nuisance of cumbering ballast, whether imported or dredged.

In common with all coaling ports, that of the Tyne is dreadfully harassed with the mass of refuse ballast; as, every ship entering the river for coal is obliged to discharge its ballast somehow and somewhere. In common, also, with all tidal rivers, it is injured by the eternal formation of sand and mud banks. It is manifestly necessary, constantly to reduce these by means of the dredging machine. It becomes, therefore, a matter of great consequence, to devise some judicious means for the depositing the dredged matter, first, where it can do no harm; and next, if possible, where it may turn to good account: for, these ballast formations or distributions, whether shamefully hoven promiscuously into the water, *as is too often the case*, or made on land, are an inconceivable nuisance, and the nightmare of all harbour masters of ports, where the exports much exceed the imports. Under such circumstances, as, on the one hand, the heaping-up of it on shore above high-water mark, is a heavy tax on trade; the carting it away, twice handling it, and mounting it up, all by dragging it *up hill*—up-hill work in every sense—being a great expense: on the other, the heaving it over-board is the destruction of the navigation, and an ultimate much greater expense; by reason that it must, all, later, be dredged up, and, especially as dredged matter is even more onerous and expensive and unprofitable to manage than dry ballast—a great boon is realizable, if, by any means that can be devised, this ballasting, wet and dry, can be gotten rid of without damage to either party.

Now in the present proposition, *universally*, as it was then proposed in the case of the Tyne in particular, the Island Breakwater requires this very material, which is in the ships coming to the port, to form it rapidly: this very material which now, in a great many cases, does mischief to the navigation. Until the Cradle, which is the frame-work of the whole structure is filled, this ballasting, both wet and dry is, of all other, the very material with which to fill it: and also, when once the work shall be completed, the continual dropping of the ballasting out of the ballast-lighters at high water, in front of the island, or for increasing its length, will be a continual advantage: the more the island grows the better; that growth being regulated by the Rules of Science.

It would seem, however, by the evidence in the Blue Book, that this sort of argument was not more likely to find favour with the Engineer of the Tyne Conservancy, than that which commended the use of the Island Screen; that Official, being of opinion, evidently, that ballasting being hoven into the water is rather a benefit, because it makes short work and saves expense. Thus, Appendix B. to Second Report, Tidal Harbour Commissioners, page 327—"Mr. Simon Danson, harbour master for the last 11 years, was bred on the Tyne, and had been to sea 43 years. Was sorry to say he had seen no improvement in the river [Tyne] of late years. Whatever additional depth had been got was owing to the dredger. He was sorry to say that too much ballast was thrown into the river. Little was done in the way of prevention: the authorities, indeed, should reform themselves before they interfered with others. What the Engineer took out at one place, he threw into another. He deposited below high-water mark, what he dredged up from the bed of the river, and a considerable portion of it was washed back again." [Evidently then they want at the Tyne, a *place* in which to deposit this dredged-up matter.] Mr. Brooks [speaks]: "It is deposited out of the action of the tide." [This, it is to be observed is an *impossibility*, although] "Lieut. Collins, of the Coast Guard, *Howden*, did speak in corroboration of the Engineer's statement." Mr. Brooks observed, that, "to land the stuff from the dredger above high-water mark, as suggested by Mr. Danson, would be very expensive and very unnecessary. None of it washed back again into the river." Alderman Dunn interposed, "And were any other person to do this, would he not be fined?" Mr. Town Councillor Lowrey,—"Why, there is a bye-law, allowing manufacturers to deposit what they choose between the jetties at 2*d*. per ton!" This is the way that the Tyne Conservancy conserve their river—the *Country's* River Tyne. "The Commissioners made a remark, implying that the stuff from the dredger should be deposited above high-water mark."

Mr. Danson (in continuation)—"Much ballast was thrown overboard from vessels within the prohibited limits" [what *are* these prohibited limits?] "and wrecks sometimes occurred in consequence. There had been a dredger in the port six or seven years: the old one was little used. They said she wasn't right. She was changed because she could not dredge alongside the quays: the new one could, and had got more depth. The old one cost £25,000 before she was laid aside as useless. In dredging, clay and gravel were brought up, as well as sand. The dredger should be used on Hepburn Shoal."

From all this it is evident that, in the Tyne, at least, ballast abounds; that the river is damaged by it; that they don't know where to put it; that while it is just the material wanted for the formation of the Island Breakwater; instead of being used for that, the dredged matter is thrown again into the water!

Mr. Brooks—"It was used there for three months, and brought up 20,000 tons of sand. The labour and cost were just thrown away." [Worse than thrown away, if the matter dredged up was all re-deposited below high-water mark: 13,000 cubic yards of it!]

Mr. Thomas Smith, river pilot, stated that—"No depth of water has been gained in the river. There was less water than there was fifty years ago. The pilots were complaining every week of the danger of taking down ships generally: the river had changed for the worse. Larger vessels came up to Dent's Hole formerly than could now come."

Mr. Joseph Straker, shipowner, &c., of North Shields—"Did not believe the river had improved, as to depth of water, in the last half century. There might be as much water as formerly, but the channel had been so narrowed that ships could not be got up, if it were not for steamers."

From these extracts from the evidence before the Tidal Harbours Commission, concerning a river of such importance as the Tyne, the necessity of intelligent supervision in all cases,—the necessity of constant dredging of such rivers,—the equal necessity of knowing what to do with the matter dredged,—as well as of knowing how to regulate and dispose of the ballast which ships coming for cargo bring with them,—will be evident.

With respect to the Tyne, it will hardly be thought strange, from what these extracts exhibit, that persons so likely to be well-informed as His Grace the Duke of Northumberland, conceive that there is very great danger of the navigation of this important river, being, at no very distant period, entirely destroyed, and that a high degree of nervousness on this subject very generally prevails. The Tyne is, in common with all rivers that debouch to the sea, undergoing a constant natural deterioration,—and this is more actively the case in tidal rivers than in others,—and the delusion entertained by the engineering class, generally, that they are able to do anything to maintain the depth of a river by any other means than by dredging, will end in our descendants finding themselves without any other inland communication with the sea but estuaries and swamps: *ex. gr.* Sandwich Flats, *iterum iterumque.* And, as in the case of the Rhine, all the dirt floated down, through its whole length, goes no further a-field than the Flats of Holland, so, in that of the Thames, all the dirt held in suspension by that river, is deposited some where, back from the mouth of the Thames, *upwards:* and in process of time, unless its formations be looked to, all these, Shoals, will become Sand Banks; the Banks, Dunes; and the space between Kent and Essex, a Delta. I don't at all mean to affirm that that would be a worse state of things than the present; but, I say, that it will be a worse state of things, as regards *navigation.* We don't want the delta, and we do want the navigation; and we shall lose it if we don't change our present system of leaving vast interests in the hands of incompetent and wrong-theoried men. The first step towards ruining our rivers, is the neglect of the Dredging Machine. It may be very expensive to lay up the dredged stuff and ballast and all that, above high-water mark. Nevertheless, that cost is the Insurance Money you pay for your river's navigation; and he who neglects to keep up this, his River Insurance, will as assuredly suffer for his neglect, as the man who finds, when the fire has destroyed all that he had, that he is reduced, from prosperity and affluence to utter and helpless ruin.

I would strongly recommend, therefore, the gentlemen of the Tyne, and those also of every other river in the kingdom, to pay up their Insurances by keeping the Dredger going in their rivers; and further, to see for and provide, beforehand, places in which, advantageously, to deposit what the dredger brings up from the bottom of their river; and not only this, but that the Country, to whom the Rivers belong—not to any Conservators with their patent rights and *iniquitous prescriptions*—see that this is INCESSANTLY done; remembering always, *as it is well that we should*, that every atom of silt or matter that is in the river between the last hour or two of the Ebb and the top of the Flood is in a constant state of deposit,—*of being deposited, that is:* and that the amount of that quantity, whatever it is, has been augmented *by the amount of matter held in suspension and brought up by the rising tide* from the ocean. *That*, is the *invariable*, eternal, *eternal Element of Deterioration*, which is at work in all our Tidal Rivers: it goes on and *it must* go on, eternally: and our only safeguard against and remedy for that, is, the patient and eternal use of the Dredger. This being the melancholy fact with which we have to contend, it is well, at once, to look out for and make provision for the reception of this eternal increase of matter, which only becomes innocent so long as we can *see* it; and that is, by its being set *above* high-water mark, whatever Mr. Brooks or any body else, with great or small reputation, persuade themselves or others to believe.

[It is satisfactory to find an entire concurrence of opinion on this point with a gentleman, with whom it ever grieves me intensely to have to differ, I mean Captain Washington, the present Hydrographer. In his Summary Report on the State of the Harbours of Scotland, under the item Steam Dredge, p. lxii., we read thus:— "In the improvement of Harbours there is no single thing that has been attended with such beneficial results as the free use of the steam dredge; in nine cases out of ten this is the grand cure for the general evil complained of; namely, want of depth * * * * nor would it be easy to point out how a greater benefit could be obtained at so small a cost." Perhaps Mr. Brooks does, and will enlighten the Hydrographer on the subject.]

No. 5.

The Sand Trap.

Captain Vetch said, a good many years since, "I have been led to form a strong opinion that none of our existing modes of construction are advisable, (either at Deal or Dover,) and that if any thing is done, we must look to new contrivances and discoveries."—*June* 1844.—*Harbours of Refuge Commission Evidence.*

"The action of the current also as a disturbing and transporting agent, may be observed on a calm day by looking over a boat's side at the passing stream, when it will be noticed that the waters move in endless convolutions, and are charged with matter; the presence of which, in this case, is the result of the friction of the brushing action of the current over the bottom. When, however, these disturbing influences are in abeyance, it loses its holding property, and the particles of matter being forced by their gravity to subside, are deposited in the first place of rest."

* * * *

A striking instance of the important part played by the current in the operation, is supplied by General Pasley, who mentions "'that he found the interior of the Royal George, silted up 29 feet, though externally there was little or no accumulation.'"—*Calver's Wave Screen*, pp. 31, 32.

"In fact, you help Nature to raise dry land out of shallow water, and form an embankment just where you require it."—*Captain Spratt, R.N., C.B., Report in the Delta of the Danube,* anno 1858, p. 13.

"As in the case of the Goodwin, the Ower, the Yarmouth, the Newcome, the *various* '*New-come*' Sands—they are all, in a sense *new*-come, as this expressive name especially implies, and they are, as surely, always growing;—so is it the case with the Smithwick Sand, in Burlington Bay. They rise to a certain height, or, as the sailors say, shoal-up; and, so long as they are exposed to the action of a sea that is too powerful for them to stand against, they are obliged to remain stationary, just so many inches, or feet, or fathoms, below high or low water mark, as the case may be: but they are all, more or less, growing and *spreading*. This being the fact, and, as we know that this growth of sand, assuming this form and that, under the surface of the sea, is very injurious to us, it will be highly advantageous, if we shall be able to devise any plan by which, while turning this element of mischief to good, we may, at the same time, abate in so far, the amount of injury done to us, by this undersea floating matter. And, if in doing this, we shall be able to convert what is now an insidious enemy into a valuable friend, we shall have gone far towards making Nature our handmaid, instead of our enemy, and, therein, of getting our work done, comparatively for nothing, instead of paying for it; and also, of making that practicable as well as possible, in many places, where, what we stand much in need of, *ex. gr.*, a work on this very Smithwick Sand, is, otherwise, entirely utterly, out of the question.

"For, how is it going to be done? It must be from two to three miles long, and it must stand in water, varying from 9 to 14 feet deep, at low-water spring tides, this very wall for which the good Mr. Straker, and Mr. Anderson, and Mr. Goodrich, and Mr. Brambles, and the Pilots, and Captain Washington, and the Tidal Harbour Commissioners, look so benevolently and longingly, this work on the Smithwick, as a boon to the seafaring man. And who is to do it for them? that is to say, *how* is it to be done, and what is it to cost?

"First, *how?* by masonry; or, 2nd, by sunken ships; or, 3rd, by caissons; or, 4th, by piling; or, 5th, by wattling?"

1st. Masonry is out of the question, notwithstanding "the ready cut and dry" phrase of young engineers, and some "eminent," that there are "no engineering difficulties." The works at Dover, have not justified that pleasant conceit, as the Treasury has found to the Country's cost.

2nd. Sunken ships? In such structures, there is no combined adhesiveness: if the Sand, on which they settle be quick at all, they go through it, and disappear as at the Goodwin. This process is, also, alarmingly extravagant: imagine three miles length of sunken ships! these notions, on any such scale of work, amount to preposterous folly, if seriously propounded as practicable.

3rd. Caissons? where is the filling of three miles length of caissons to come from?

4th. By piling? *wood* piling is not *impossible*, but it is impracticable, by reason of its great cost, and of the necessity of having good weather to work in. Besides, piling could not be set or made to stand so close as to be impervious to the entrance of the sea, raging and working itself through the interstice; and thence, infinite mischief. If wood piling would cost much, iron screw piling would be still more onerous, nor would the system answer for such a work as a breakwater; the object of which is, shelter and safe lying behind.

5th. Wattling? this though extensively in use on the Holland and Belgian coasts, is not available in deep water, and could not be at all applied, until the level of the sand were brought up to the height of low-water mark spring tides. It is used for the formation and defence of fore-shore, by the accumulation of sand, which it encourages; for which purpose it is laid down in parallels to the water's edge; the parallels rising gradually as they recede from the sea at convenient height, one above, and at convenient distances, one from, another. The process is, at any rate, expensive in labour, very tedious, and would be neither so effective, certain, or lasting, as the proposed plan.*

* There is still another system of work invented by Mr. Calver, R.N. It is very taking, but I don't think it applicable in all cases. Even if the theory be correct, the object, in the present instance, would not be realized; for if it be to still the Sea, it is no less to be a barrier to prevent the sand from the sea getting over the Sand; and so, in time, destroying that Anchorage which is now so available and so unspeakably valuable; Mr. Calver's open-work Piling would not do this. And wherever the object is to keep the anchorage clear, and the shore happen to be infested by sand, it would be absolutely mischievous. It would promote the formation, by multiplying the eddies, rather than help to disperse it. With great respect for Mr. Calver, and with much thanks to him for his very useful publication, I think that the only difference between him and myself is, that he proposes one sort of Island, I, another. *I* think that mine is a good island, while I

"It is believed therefore, that all these systems of construction being impracticable, in this or such instances, and ruinously expensive withal, a system of building under water, *incredibly cheap*, such as shall, *at once*, realize the object in view by being laid down, and that, in such *mode* of construction, so that it shall entrap and hold all the sand that every succeeding tide shall bring up or down, and every succeeding wave carry into it, will entirely command the necessities of the case; and, in so doing, not only show how the wants of Yarmouth, the mouth of the Thames, the Goodwin Sands, and a hundred other places, are to be met, with ease, economy, and dispatch, but, also that the navigation in all the neighbourhoods of such works, will be greatly benefited by reason of a considerable absorption of the now-floating mischief, by which our seas are infested.

"The case is as clear as can be, and may thus be stated to become intelligible to all. If Sands will form by the mere force of gravitation, in the infinite eddy of sea and tidal water in motion and in opposition, much more will the sand, form and consolidate, in Traps prepared for its reception and, once entered, protection against the power that now disperses seven-eighths of it, while the one-eight of it, or even one sixteenth of it, is depositing.

"It is to be remembered, with respect to the amount of this deposit, that the Dunes of the opposite coast, extending from much below Dunkirk—*i. e.*, the Kirk in the Dunes—all along northwards to the mouth of the Elbe, are nothing more than the eternal accretions of sea sand, laid by the action of the water, first along shore, between low and high water mark, and then, by the action of the wind, lifted off and wraithed along the shore to their ultimate assumption of the forms in which we see them now, thus self-developed, as it were. If the records of our own seas,

---

think that his is inferior to it, more expensive, and less lasting; and, as an Island Screen, not so efficacious. My island is as little obnoxious to the tide stream as his: a point much in favour of the Island Plan of Treatment.

therefore, with their ever-occurring "newcomes," left any room to doubt of the *eternal* action of this insidious enemy, those over-sea Dunes of the opposite coast, not less than the Downs, as *we* call them, opposite Deal, hardly detached from our own, make it evident that there is abundance of matter to deal with, and abundance of the element of mischief, if possible, to abate.

"I think that there can be no reasonable room to doubt but that a *systematic* attention to this subject would produce a great improvement in the condition of our Coasts, as to the navigation of them; nor that the increase of land, and recovery of lost area, might be so profitable as to repay the cost of the provident care-taking thus hinted at. The site of the cathedral of the old See of Dunwich is now seven miles at sea. At all events, apart from all that, here is a substantive good to be effected in meeting wisely and rationally, a frightful and crying want. It is a good to be realized by taking advantage of the Elements of Nature; and, by a slight assistance from Art, turning her powers and her properties to man's account, and to the nation's profit.

[It is wretched to think of what has been lost to the Country in consequence of ancient neglects. The mother of Mr. Maggs, publisher at Southwold, recollected many miles of Land and Foresting washed away, in her time, in this very locality. Here was the material out of which, with the sea accumulations, those frightful banks of sand that assist to make our coasts the graves of British seamen, have been formed.]

"This System of Sand Trap is not a creation of the brain, for any such purpose as is here spoken of. This Work was laid out for an altogether different object. It was intended as a Framework, in and by which, to confine and protect, from the action of the sea, *ad interim*, the successive layers of horizontal masses of concrete with which it was proposed to accumulate the sea-wall of that now well-known, because long-expected yet non-existent, work, the Harbour of Refuge of Dover.

"The contemplation of the Model, in conjunction with the

knowledge of the name of the not-far-off-lying *New*-come Sand, suggested the idea of the Horizontal Sand Trap; by means of which, as laid upon the crest of a sand, to convert the Under-sea Breakwater, which every shoal is to the anchorage behind it, into an Over-sea Breakwater; taking advantage of the already manifested tendency of the sand to form in the locality where shoals are found, by affording it a cellular shelter in which to accumulate.*

"The next thing to be done was to look out for a suitable locality on which to place such a work. The Smithwick Sand, which had long been in the eye of sea-faring men as such, was soon called to my notice; and I proposed, in a pamphlet on harbours of refuge, which I published now 13 years since, that the Smithwick Sand should be the foundation for the Breakwater, that, on this system of building, should make Burlington Bay a Harbour of Refuge on the Eastern Coast.

"Since that time, as well as before it, the proposition having commanded the approval of men, well-qualified from professional experience to pronounce an opinion, it is presented as the ready and economical means of realizing such a work as the Smithwick Sand must have, to enable it to do that for the Mariner and the Nation, which the experience of able and earnest men has shewn to be its Vocation; and, it is hoped, under the intelligence of the Honourable House and its authority, its Destiny."

[It may be observed in general, that when this work would be placed on a Sand, it would always be used as a Sand Trap; but if it had to be applied as a Breakwater Screen, in front of a river, it would be right to take advantage of all means that might present themselves of expediting the process; as, for example, in the use of ballasting, dry or wet, from the shipping or from the

---

* *See* mottoes to this Paper, and Capt. Spratt's Report, Delta of the Danube.

dredging of the river: or, otherwise, to fill it up by concrete, according to the opportunities of the place or the exigencies of the occasion.

Thus, it is pertinent to observe here that I was informed by Mr. T. Baker Cresswell, M.P. for Northumberland, when he saw the model of this work in 1845, that the occasion of the failure of the works at Warkworth, was, the heaving of the Rubble into the sea, without defending the accumulation by any mould or frame. Hence, it was dispersed by the violence of the sea. Had the Warkworth Engineer filled his rubble into a horizontal cradle, the work would have succeeded—at least, so thought the Hon. Member—and the £75,000 which was there lost have been saved.]

No. 6.

Cost—Small Cost, Many Breakwaters—Prodigious Outlay, None Possible.

"The difficulty and expense of constructing such harbours have, in short, been the chief impediment to forming them, and some more economical plan is loudly called for: something of so inexpensive character as would enable the number of refuge harbours round our sea-board to be multiplied twenty fold."—*Calver*, p. 68.

"They should be inexpensive, so as to admit of their being established in every situation where shelter is required, and where the nature of the frontage will admit of it being allowed."—*Calver, Wave Screen*, p 44.

"And then it is stated by the Chancellor of the Exchequer, that to make these harbours of stone, would cost six millions of money; and as 250 harbours are reported to be required for the protection of Great Britain, it would be impossible, of course, to do that."

(2134.) "It has been stated by the Returns. That was stated in the House of Commons to be the number that was required. But there is no doubt, that if you take every inlet, that number is required; and particularly now, in the event of a war, with steamers, you would require every inlet to be navigable and to be a protection."—*Evidence of Admiral Taylor, C.B., Select Committee on Shipwrecks.*

THE next great advantage derivable from the adoption of these Island Screens, will be evidenced by Captain Bullock's testimony. (Report Select Committee on Shipwrecks, page 294. Question 4394.) "Do you consider that a harbour of refuge, built by the Government, would be most desirable for the shipping, or that there should be something more in the character of breakwaters?"—"I think a harbour could be safely constructed, even a permanent harbour; but they are of so expensive a description, I think it would be worth the attention of the Committee, whether they should not recommend a breakwater: a breakwater could be constructed, I suppose, for at least one-twentieth of the expense.

I could construct a breakwater for £100,000, where it would cost millions for an harbour. If the Committee will receive as a principle, that a lee can be obtained by holding a body of a particular construction across a wave, *we* can go on so far as to say, it would be possible to keep it in its position. We know well, that, where a line-of-battle ship at anchor tends to windward, say in the Downs, one can lie beautifully under her lee: the difficulty is how to fix it, permanently. It is my opinion, that that *could* be done; because, if an anchor of a given weight will not do, a double or a treble one will."

(4395.) "Do you apply that, floating or fixed?"—"Floating: then they can be arranged as you please: it is difficult to adapt screens. I have seen the plan of Mr. Cubitt; but, that screening of Dover Bay is, I think, impracticable: *he* says not, and he is a clever engineer."

(4401.) "Suppose piers, [island piers,] could be constructed such as are described by Mr. Cubitt, of a solid nature, do you not think that they would answer better than a floating breakwater?"—"Yes, it is the expense which forms the objection; for, if the breakwater succeeded, we could have twenty harbours for one."

It here appears, that Captain Bullock thinks it would be a good outlay, even for a floating breakwater, to spend £100,000, in any favourable place; and, *if possible*, to have Mr. Cubitt's plan carried out, at the cost of, say, some £700,000 for a mile length of work; of £400,000 for half a mile; of £250,000 for a quarter of a mile; of £150,000 for 290 yards. (*See* Answer to Question 4059, p. 262—*ibid.*) The whole sum (4046), was £2,000,000.

But the plan herewith recommended to the Committee, which also Captain Bullock has himself seen and commended—and the man who has erected a Beacon on the Goodwin Sands (*vide* p. 287—*ibid*), must be supposed to know something of possibilities and practicabilities in such things, will take *all cases* in hand; will not only "screen" as by floating breakwater, but make "a solid work" in front of every barred river, at a cost greatly less than Captain Bullock supposes floating works can be set up for. When this is done, then we shall have his words, and his more than hope, verified,—"Yes, it is the expense which forms the objection; for, if a breakwater succeeded, we could have twenty harbours for one."

When it occurred to me first to propose the plan or mode of building, here brought forward, for the erection of sea wall, it was not with any idea of the cheapness of the plan, but on account of its facility of erection, and of its being the best then devisable, or at any rate, spoken of.

It happens, however, that what is best, is also cheapest; and that things, utterly impracticable otherwise, are attainable by this method of construction. The plan was not only recommended, and avouched by high practical authority, as an Engineering Device, but it could speak for itself as to Price.

When, in the year 1846, I printed a pamphlet on Harbours of Refuge, Colonel Sykes said—"Unless you put out an Estimate, shewing that the work you propose for Dover *can* be done for such and such money, as you allege, people will not believe it to be possible." Accordingly, in self-defence, as well as for general information, the accompanying document, marked No. 7, headed "Harbour of Refuge. Estimate under £500,000, of the Cost for Material, Construction, &c., of a Breakwater in Dover Bay," was printed by me, to go along with the pamphlet.

The allegation, for example, at that time by the Harbour of Refuge Commission, was, in the case of Dover, (for which case particularly the Pamphlet and Estimate were printed,) that they would have a certain work, done in a certain way, to cost the Country £2,500,000. My assertion to them was—"While you have stated £2,500,000, you know you mean it to cost £5,000,000. I know that you won't accomplish it for ten millions, and there is a greater probability of its costing twenty millions than ten, even if you can do it at all on the plan you propose, which is doubtful; and you will, besides, be 20 years in doing it." The results have verified these predictions; £650,000 has been spent for doing less than one-fourteenth of the mural area of the proposed work.*

By the plan I propose, the same work would have cost under £500,000. It is practicable as well as, and instead of being, barely possible: it will be useable in one year, and finished in two. The Estimate was the proof of the first part of this proposition; the Model, now at the United Service Institution, that of the last.

The reason alleged for not entertaining my suggestions was, that it was "too late." But it was *not* "too late:" for they did not commence operations for two years afterwards; not venturing, as I was told with some rancour by an interested party, too boldly to disregard the statements in that pamphlet; and, unfortunately, my predictions with respect to what would occur at Dover, have been but too painfully realized. After 12 years, they have only completed 300 yards of the easiest part of the work; in eight years more they are to have completed 600 yards of a plan which is to cover a length of 3,300 yards, *i. e.*, less than a fifth of the work in 20 years!

It has been generally stated, that the works up to this time have cost £3,000,000; which however, I don't at all believe; above a million has been long fiddled-away there with a Scouring

---

* This is not exactly the case. This was stated in the House, and I wrote under that authority. £650,000 is the Estimate for the South Pier, but £400,000 it seems, *have* been spent. What has been spent besides in making the Digue and Fending Works to the Esplanade and Waterloo Crescent, who can say? That is, perhaps, to be ferreted out of the accounts of the Dover Harbour Commissioners, and not those of the Admiralty Pier: though as the folly of the latter work necessitated the doing of the former, it is, therefore, properly, *all* to be reckoned as one amount. We shall see, later, according to what estimate of probability the £10,000,000 may become £20,000,000. Sir Christopher Wren and the Paris slab! *Cavendo tutus.* It is very desirable that a complete account of the whole expenditure in Dover, its Waterworks and Harbour, should be put out. Exaggeration would then cease to be mischievous.

Toy; and it is more than probable that another million has been expended in these last years, on what is now designated the Admiralty Pier, and, IN REPAIRING THE MISCHIEF CAUSED BY THE PROJECTION OF THAT MOST INJUDICIOUS WORK. But, for the HALF of that million, might now, already, 12 years ago, have been completed, in a much more efficient manner than would be the the case even if their own plan were carried out to the letter of their promise, a harbour strictly according with the form of that proposed by them."*

[I did not venture to propose any alteration in the form of their Plan. I only suggested as a remedy against the feared evil of the silting up of the proposed close harbour, a system of tunnels through the East and West Piers of the work: a precaution not only in accordance with the suggestions of modern experience, but with the wisdom of the ancients. Sir John Rennie says, "I believe the moles of Tyre were made in that way. Certainly the moles of Romans at Pooruoli, at Porto Guilio, in the Bay of Baia, also at Misenum, which was the station of Augustus's fleet, and the ancient Ports of Antium, Ostia, and Ancona were made so. They were all constructed on arches, and their object was to produce sufficient circulation of currents through the arches, to prevent any silting up, by the water being just-sufficiently agitated to keep open the Harbour when made, and, at the same time, to have sufficient solidity to break the sea:

---

* I *believe* that Mr. Cubitt's Plan of Harbour for Dover was what ought to have been adopted. See page 261, Select Committee on Shipwrecks.

Had that been done, none of the cost of the Dover Esplanade Digue would have been incurred, and it was *a most unpardonable* thing, and reflects infinite discredit on the authorities who determined, to run out the Piers *first;* because, they were warned and *fore*warned of what would occur. But they chose *to do;* and lo! the result. In order to save the Esplanade and Waterloo Crescent from toppling over into the sea, as did the Ardglass Lighthouse, the Digue became a necessary work. This it is to defy warnings and to flout at those whose "business it is 'to mind their own business.'"

a most ingenious and scientific principle." Answer to Question 6047, Select Committee on Shipwrecks.

This appears to be a true principle. Whenever it becomes necessary to lay a Breakwater across a tide way, it ought to have a system of tunnel through it, below low-water mark.]

"And in what a condition is now, Dover? awaiting harbour that was wanted in such a hurry 14 years ago! that place, of which the present Hydrographer in his evidence already quoted, and in which he embodied and expressed the feelings of a vast body of the intelligent of the day, answered (Q. 1578:—Are there not facilities for forming a harbour of refuge at Hartlepool?)—"Yes, Hartlepool is a very fair harbour. Hartlepool at present is one of the best harbours we have there; and I think even under Flamborough Head it might be done. Flamborough Head affords a fine natural breakwater, *but the place where a Harbour of Refuge seems to me, most wanted, is Dover.*" Q. 1579. "Why do you consider Dover a spot where a Harbour of Refuge is particularly wanted?"—"Because so many thousands of vessels pass there."—*Captain Washington's Evidence.* Select Committee on Shipwrecks.

"Thus spoke Captain Washington in May 1843.* Two years later it was determined upon, there *should be* a harbour there, and consent of Parliament obtained. And now after 14 years, there is no sign of the harbour of refuge but the pier which should never have been built.† Why this delay in doing what has been ordered? My theory of the dawdling about it is, that the Authorities have found out the impracticability of the whole affair, as proposed. They don't choose, *now*, to do what was pointed out to them as reasonable and practicable; and meaning to forego one main feature of their plan, they are dawdling away 20 years in

* For the evidence which went to the formation of the public mind as to the necessity for a harbour of refuge in Dover Bay, *see* Captain Washington's Evidence, *ibid.* Questions 1579—1599 inclusive, p. 102.

† Cubitt's Dover Evidence, 2068—9.

constructing 600 yards of it; since it may be reasonably presumed that all those who concurred in the stupendous folly of decreeing the erection of perpendicular stone walls, on a soft foundation, in 7 fathoms of water, where the tide rises 20 feet also in the open sea, will be in their graves, and nobody of them will be living to have to endure the disgrace of an exploded folly. This may be a very audacious supposition. It is. Be it so! Let then the public be informed WHY a work that was to have been commenced *immediately* and finished *without delay*, is thus wearily dragging on its stunted length. The *necessity* for a War and Merchant-Fleet Harbour at Dover was paramount, they *said;* and yet 20 years are to be exhausted in doing but the first fifth of it! [I beg most markedly to exculpate any members of the Harbour of Refuge Commission who voted for the long slope system of building from this censure. Sir Howard Douglas, then M.P., protested against the method proposed, in a separate Parliamentary Paper.]

"The history of the matter appears to me to be this. They would not be taught, and the end of it may be, for aught they care, that our fleet may be snapped up some of these days, in order that the great reputation of some very small men may be saved. It is time for Parliament to look to the matter; for if the pressing reasons which were given 14 years ago were worth anything then, they are worth ten times as much now. If, on the other hand, the whole affair was a delusion, there has been frightful waste of money, of time, of talk. And upon whom is to fall the responsibility? or are these squanderings and wastings, and fanciful imaginations of "eminent men," to be fed and indulged and revelled in with impunity? *

---

* Thus, in one instance, that "very sanguine Engineer" Mr. Rendle, will have a high sill across the opening of a harbour at the bottom, as if to make the harbour silt up to *that* level at all events (Holyhead); in another instance, he sets the sill of his dock gates below the ordinary level of the river, so that it is impossible to open the gates when needed; and for such eccentricities and vagaries John Bull, has to pay in *Life* as well as in Money. It is since these Original Papers

"Now with respect to Dover, its wants, and the wants of other places equally with it, and the cost of the requisite sufficient works, I have never changed my note. What I said 14 years ago I repeat without the least change, and will abide by now. I said first, that the work, as proposed for Dover Roads, was impracticable; and 2nd, that, for the intended £5,000,000 that was to be devoted to Dover, Works could be constructed at Dover, Yarmouth, Cromer, Bridlington, and the Tyne, and for less money. I still maintain the same. There are engineers of note, if that be necessary, who will avouch the Plan, and the Estimate that is herewith sent, and who will avouch the cost of any and of all the works. And I feel that I have a right to add this further, that, instead of the abortive attempt at Dover, which, besides being a cruel waste of public money, is a scandal to the engineering reputation of the kingdom, being all that we have for our wasted treasure, there would have been, in the providence of God, now living, of men drowned in wrecks at sea, 7000 human beings; as many ships in existence which have now perished, and, together with them, as many millions of private property of the citizens of this empire, as would have sufficed to make all the works together that I have above specified. All this it is to be remembered, *is not declamation;* it is matter, all, of hard fact and proof! And I say, that the lives of these men, the prey of incapacity and folly, may be laid at the door of, and the cry of their widows and orphans

---

were forwarded to Admiral Walcott, that I have, by accident, seen a memorandum by (the then) Captain F. W. Beechey, R N., F.R.S., on the Artificial Harbours of the Irish Sea, addressed to the then Hydrographer to the Admiralty, the late Admiral Sir Thomas Beaufort. Plenty of vagaries of eminent and "very sanguine engineers" there to be seen or guessed at. (Re-printed here, p. 81.) It will quite illustrate this sentence; and it must be remembered that, whenever money is spent upon the Coast, as it is done under the *controul* of the Admiralty Engineers, it is, therefore, under the seal of the Admiralty authority that all these vagaries are perpetrated. A responsibility that it may be very hard upon the Admiralty to have to bear, but then the remedy lies with themselves Let all future Lords of the Admiralty, after the example of Sir John Pakington, use their own Common Sense, and be guided by Experience and Evidence.

may ascend against, those and all men, who having the means to command, and *having* ASSUMED *the responsibility of office* which has required them to ordain, have callously or carelessly been parties to such woeful wicked waste of this noble element of our Country's greatness."

Let the Reader here, *read*, and *re-read*, and ponder upon, the Epigraphs on the Title Page of this Pamphlet! They declare frightful and disgraceful truths and facts.

No. 7.*

The Estimate.

Q. 3894. "When did the works at Dover commence?"—"About ten years since, I think."—p. 201. Select Committee on Harbour of Refuge.

Q. 3895. "Was it known at the time it was decided to make these works at Dover, that it would take about half a century to complete them?"—"I do not suppose it was: I do not think, at the time it was decided, that any correct idea was formed as to the mode in which it could be done."—Mr. Walker's Evidence.

"FOR WHICH OF YOU, INTENDING TO BUILD A TOWER, SETTETH NOT DOWN FIRST AND COUNTETH THE COST, WHETHER HE HATH SUFFICIENT TO FINISH IT."

THIS Estimate is reprinted now, for the same reason that it was put forth in the first place; viz., to prove to those who chose wilfully to deny possibilities, that he who affirms, has proof to shew that his promises do not mean two for one, nor their realisation four for one, on delusive expectations delusively held out. It is an estimate *in detail*, and not in the lump. The prices, however, were those of 1845, and every thing has much risen in price since that, as well Labour as Material. Once, in an illustrious presence, the assertion was made, "It can be done for no such money;" on which, the noble Duke took up the Estimate, (the original M.S. from which this Paper was printed) and said, "Why not? so many feet of timber at so much per foot—rule-of-three sum!" whereupon the opposer was "shut up." I mean

---

* The only part of this Paper, No. 7, which was portion of Document No. 2, sent to Admiral Walcott, and presented to the Royal Harbour Commissioner, is—The Estimate.

it now to shut up a good many others, who make the same pretended impossibilities the plea for spending millions where hundred-thousands should suffice.

I desire, therefore, to put it on *record*, once and again, that in the year 1845, it was pointed out to the Authorities how, for £500,000, the works at Dover could have been completed in two years, on a plan avouched by an Engineer Officer, a man well experienced in under-water constructions in the River St. Lawrence; Colonel Jackson, of the Royal Staff Corps: and to let the same public know that, after 12 years labour, and an outlay of as much as would have sufficed for a Breakwater, about 1-14th of the whole work is done; so much or so little as suffices *not* to shelter a cock-boat of a packet when it arrives at low water with the wind on-shore. Verily, this *is* a splendid result after 12 years work, and the previous million spent in their absurd tool, their Scouring Toy.

His Grace the Lord Warden of the Cinque Ports propounded one rule-of-three sum to Mr. Walker in my presence.—"If one foot of timber cost 1*s*., what will * * cost?" I beg to propose another which is not quite so simple, being of Two Statements; though I apprehend John Bull will understand it quite sufficiently. I commend it to Mr. Bull's representatives, The Commons House of Parliament.

First Statement: If it take 12 years to do 300 yards of work adjoining the shore, with railroad appliances and land carriage available, and, in water not beyond five fathoms depth, how much will it take to do the whole work which is to be 3300 yards in length, of which nearly 3000 remain to be done, and of these 3000, more than 2000 are out at sea, and the greater part in a depth of seven fathoms at low-water?

Second Statement: If it cost £400 a foot in-shore, without having to dig-out deep holes of soft-chalk and rotten foundation, how much will cover that same contingency, when aggravated by the pressure of deeper water, and out of holes the depth (of bottom) of which, no man can calculate?

Therefore I repeat, *Cavendo tutus!* Stop all these works, and begin to build upon a rational plan. Better late than never! You will remember, Gentlemen Commissioners, all You whom it may concern, You were told this *before* you commenced—You were warned of that which would almost certainly occur, and You derided the warner, and boldly promised "no engineering difficulties;" and as boldly put out your hands to scoop out the thousands from the Treasury. And You despised the rough and ready, and cheap and yet compact and *continuous** plan, by which, as below, you would have worked in safety and with rapidity; and withal, in *safe disregard* of the evil contingencies of *proved* soft bottom. Now You have the experience of one whole year's work of digging out one hole of soft chalk, do You still hold to the bold experiment of "going into" what may chance you in the deep sea of seven fathoms? And if *You* are so bold, will the Country be "so unheeding of reasonable probabilities" as to allow you to dare the folly? Come forward then now, and give an account for yourselves. You have spent the money, and where is the shelter? Was not one Packet not three months since obliged to seek another Port at midnight, because, after an outlay of a million and a half, the last half by You, Commissioners and Commissioner-Engineer, there is not security of entrance for a thing that draws so little water as a channel steam-packet?!

Good Colonel Sykes, in his simplicity, wrote to me, "You have at all events done this good (in printing the Pamphlet); you have prevented five millions being thrown into the sea." Yes, so we thought. But within a month or so after the Gallant Colonel so wrote, the beginning of the "throwing into the sea" commenced. And they have been diligently at work, at that pleasant operation, ever since: and lo, the result! And they are not a bit abashed!

* "The grand desideratum in harbour works, which is the want of continuity in the structure, could be supplied. By the use of long logs of timber, carefully bolted together, a new element of strength is obviously obtained."—*Mr. Stevenson.*

Extravagancies and blunderings are going on still in twenty other places. There they are, blundering away at Newcastle; and a Bar of 3 feet at "the perfect harbour of Sunderland" affords promise of a wonderful field of Waste and Folly,—and Hartlepool Roads, to be walled at a cost of £800,000 into a mud-pool,—and the Filey Folly at £850,000, and more *ad libitum*, &c. &c.—to some lucky "Eminent Engineers," with £5 per cent. premium upon all that they can spend. And, the longer the jobs last the better; for so, opportunities occur of increasing £500,000 to £600,000 (see present state of the Tyne); with the prospect of increasing £600,000 to £800,000 (see present state of the Tyne, *iterum iterumque!*):—all which, altering and changing of plan, shewing clearly, that the engineers so-called, did not know their own business, any further than to make plans which should require changing; and so, give them the occasion of pulling down to begin again; and herein, of course, a wonderful opportunity for increasing the tale of per centage. With a percentage on £10,000,000, that Dover Job might go a good way towards building a palace in Great George Street, Westminster.

I venture to think that a system of payment, which is neither more nor less than a premium upon extravagance, should be altered: it is much to be desired that that respectable memberhood, the Civil Engineers' Association, should themselves come to some resolution on the subject of a creditable system of remuneration.

THE ESTIMATE.

TABLE I.—Showing the Quantity and Cost of Timber in each Layer, and of Timber and Concrete to be used in each Mass, or Portion of Fifty Yards, of the proposed Breakwater; the Masses varying with the Soundings.

Soundings at low water, as marked in the Charts.	Do. high water.	Number of Layer.	Number of cubic feet of balk in each layer.	Cost of each layer. £	s.	d.	Vide C. I. Table II.	Cubic feet and cubic yds. of timber in each mass.	Cost of timber in each mass. £	s.	d.
			feet ft. in. ft. in.	£	s.	d.			£	s.	d.
		1	320+ 80 =400	20	0	0		The upper layer.			
		2	320+ 84 =404	20	4	0		(Six feet above			
		3	320+ 88 =408	20	8	0		H. W. M.)			
		4	320+ 92 =412	20	12	0					
		5	320+ 96 =416	20	16	0					
		6	320+100 =420	21	0	0					
	1	7	320+104 =424	21	4	0		High water mark.			
	2	8	320+108 =428	21	8	0		(Two feet above			
	3	9	320+112 =432	21	12	0		L. W. M.)			
	4	10	320+116 =436	21	16	0					
	5	11	320+120 =440	22	0	0					
	6	12	320+124 =444	22	4	0					
	7	13	320+128 =448	22	8	0					
	8	14	320+132 =452	22	12	0					
	9	15	320+136 =556	22	16	0					
	10	16	320+140 =460	23	0	0		Cub. ft. Cub. yds.			
	11	17	320+144 =464	23	4	0	(a) 2	6880 — 254	344	0	0
	12	18	320+148 =468	23	8	0					
	13	19	320+152 =472	23	12	0					
	14	20	320+156 =476	23	16	0					
	15	21	320+160 =480	24	0	0					
	16	22	320+164 =484	24	4	0					
	17	23	320+168 =488	24	8	0					
	18	24	320+172 =492	24	12	0					
	19	25	320+176 =496	24	16	0					
	20	26	320+180·4=500·4	25	0	4	(b) 1	11700·4— 433	385	0	4
1	21	27	320+184·8=504·8	25	4	8		Low water mark.			
2	22	28	320+189 =509	25	9	0					
3	23	29	320+193·4=513·4	25	13	4					
4	24	30	320+197·8=517·8	25	17	8					
5	25	31	320+202 =522	26	2	0	(c) 1	14267· — 528	713	7	0
6	26	32	320+206·4=526·4	26	6	4	(d) 1	14793·4— 547	739	13	4
7	27	33	320+210·8=530·8	26	10	8					
8	28	34	320+215 =535	26	15	0	(e) 1	15859· — 587	792	19	0
9	29	35	320+219·4=539·4	26	19	4					
10	30	36	320+223·8=543·8	27	3	8					
11	31	37	320+228 =548	27	8	0					
12	32	38	320+232·4=552·4	27	12	4	(f) 1	18042·4— 668	902	2	4
13	33	39	320+236·8=556·8	27	16	8	(g) 1	18599· — 668	929	19	0
14	34	40	320+241 =561	28	1	0					
15	35	41	320+245·4=565·4	28	5	4					

(Continued)

TABLE I.—Showing the Quantity and Cost of Timber in each layer, &c. (*continued*).

Soundings.		Number of Layer.	Number of cubic feet of balk in each layer.	Cost of each layer.			Vide C, I. Table II.	Cubic feet and cubic yds. of timber in each mass.	Cost of timber in each mass.		
			feet ft. in. ft. in.	£	s.	d.			£	s.	d.
16	36	42	320+249·8=569·8	28	9	8	(*h*) 2	20225· — 751	1014	15	0
17	37	43	320+254 =574	28	14	0					
18	38	44	320+258·4=578·4	28	1	4					
19	39	45	329+262·8=582·8	28	1	8					
20	40	46	320+267 =587	29	7	0	(*i*) 1	22617· — 837	1130	17	0
21	41	47	310+271·4=591·4	19	2	9					
22	42	48	320+275·8=595·8	29	15	8					
23	43	49	320+280 =600	30	0	0	(*k*) 1	24404· — 903	1220	4	0
24	44	50	320+284·4=604·4	30	4	4	(*l*) 1	25008·4— 926	1250	8	4
25	45	51	323+288 8=608·8	30	8	3					
26	46	52	320+293 =613	30	13	0	(*m*) 1	22630· — 971	1311	10	0
27	47	53	320+297·4=617·4	30	17	4					
28	48	54	320+301·8=621·8	31	1	8	(*n*) 2	27469· —1017	1373	9	0
29	49	55	320+306 =626	31	6	0					
30	50	56	320+310·4=630·4	31	10	4					
31	51	57	329+314·8=634·8	31	14	8					
32	52	58	320+319 =639	31	19	0					
33	53	59	320+323·4=643·4	32	3	4					
34	54	60	320+327·8=647·8	32	7	8					
35	55	61	320+332 =652	32	12	0					
36	56	62	320+336·4=656·4	32	16	4	(*o*) 2	32598·4—1207	1629	18	4
37	57	63	320+340·8=660·8	33	0	8					
38	58	64	320+345 =665	33	5	0					
39	59	65	320+349·4=669·4	33	9	4					
40	60	66	320+353·8=673·8	33	13	8	(*p*) 6	35267 —1306	1763	7	0
41	61	67	320+358 =678	33	18	0	(*q*) 14	35945 —1331	1797	5	0
42	62	68	320+362·4=682·4	34	2	4	(*r*) 9	36627·4—1356	1831	7	4
43	63	69	320+366·8=686·8	34	6	8	(*s*) 9	37314· —1382	1865	14	0
44	64	70	320=371 =691	34	11	0	(*t*) 4	38005 —1407	1900	5	4
45	65	71	320+375·4=695·4	34	15	4	(*u*) 5	38700·4—1433	1935	0	4
46	66	72	320+379·8=699·8	34	19	8	(*v*) 4	39400· —1459	1970	0	0

The basis of the above given calculation is, that the best Baltic Timber can be obtained by the Government at the rate of One Shilling per foot. The market price of Timber is now (Nov. 1845) about £4 15*s*. per load.

	£ s. d.		£ s. d.	
Baltic, best Crown .	1 17 0	Average per load (say)	1 15 0	£4 15 0
2nd ditto	1 15 0			
3rd ditto	1 12 0			
Freight, per load			0 15 0	
Loading and carriage			0 5 0	
Add duty			1 6 6	
Merchant's profit			0 13 6	

From which, deducting duty and profit, the price per load remains £2 15*s*., or 1*s*. 1¼*d*. per foot. But from information obtained from a Norwegian Captain in Dover, the price per load is £1 10*s*, which, with freight, and loading, and carriage, makes just 1*s*. per foot.

The cost of Concrete is estimated at Two Shillings per cubic yard.

TABLE II.—Showing the Cost of the Timber and Concrete to be used in the Whole Work, (exclusive of the the Tunnels,) with the Parapets, and also for contingent Extra Width of, and Solidity in, the Mass.

Number of each Mass.	Depth in Feet, and the Number of each Mass, as marked in the Plan.	Cost of any Mass Timber.	Aggregate Cost of all the Masses.	Cost of any Mass Concrete.	Aggregate Cost of all the Masses.	Superficial feet in the vertical face of any Mass.	Ditto in ditto as in 69 Masses.
		£ s. d.	£ s. d.	£ s.	£ s.		
a 2	16 1·67	344 0 0	688 0 0	472.2$\frac{21}{27}$	944. 5$\frac{10}{27}$	2400	4800
b 1	26 2	585 0 4	585 0 4	910.0$\frac{23}{27}$	910. $\frac{23}{27}$	3900	3900
c 1	31 3	713 7 0	713 7 0	1171.6$\frac{1}{27}$	1171. 6$\frac{1}{27}$	4650	4650
d 1	32 66	739 13 4	739 13 4	1229.8$\frac{13}{27}$	1229. 8$\frac{13}{27}$	4800	4800
e 1	34 4	792 9 0	792 9 2	1346.5$\frac{11}{27}$	1346. 5$\frac{11}{27}$	5100	5100
f 1	88 5	902 2 4	902 2 4	1595.4$\frac{25}{27}$	1595. 4$\frac{25}{27}$	5700	5700
g 1	39 65	929 19 0	929 19 0	1660.9$\frac{8}{27}$	1660. 9$\frac{8}{27}$	5850	5850
h 2	42 6·64	1014 15 0	2029 10 0	1864.9$\frac{7}{27}$	3728.18$\frac{14}{27}$	6300	12600
i 1	46 7	1130 17 0	1130 17 0	2154.8$\frac{2}{27}$	2154. 8$\frac{12}{27}$	6900	6900
k 1	49 8	1220 4 0	1220 4 0	2385.7$\frac{7}{27}$	2385. 7$\frac{17}{27}$	7350	7350
l 1	50 63	1250 8 4	1250 8 4	2465.2$\frac{11}{27}$	2465. 2$\frac{18}{27}$	7500	7500
m 1	52 9	1311 10 0	1311 10 0	2628.1$\frac{8}{27}$	2628. 1$\frac{8}{27}$	7800	7800
n 2	54 10·69	1373 9 0	2746 18 0	2796.1$\frac{14}{27}$	5592. 3$\frac{1}{27}$	8100	16200
o 2	62 11·12	1629 18 4	3259 16 8	3519.2$\frac{16}{27}$	7038. 5$\frac{5}{27}$	9300	18600
p 6	66 13. 14. 15. 16. 61. 62	1763 7 0	10580 2 0	3911.6$\frac{26}{27}$	23468. 1$\frac{26}{27}$	9900	59400
q 14	67 {17. 18. 19. 38. 39. 40. 41. 42. 45. 46. 47. 48. 59. 60.	1797 5 0	25161 10 0	4012.9$\frac{22}{27}$	56174.17$\frac{23}{27}$	10050	140700
r 9	68 20. 23. 34. 23. 26. 35. 51. 52. 58.	1831 7 4	16482 6 0	4115.5$\frac{22}{27}$	37037. 9$\frac{8}{27}$	10200	91800
s 9	69 21. 27. 36. 37. 43. 44. 49. 50. 68.	1865 14 0	16791 6 0	4219.3$\frac{25}{27}$	37972.15$\frac{9}{27}$	10350	93150
t 4	70 22. 21. 34. 53.	1900 5 4	7601 0 0	4324.3$\frac{6}{27}$	17297. 0$\frac{20}{27}$	10500	42000
u 5	71 29. 32. 33. 54. 57.	1935 0 4	9675 1 8	4430.9$\frac{7}{27}$	22153. 8$\frac{4}{27}$	10650	53250
v 4	72 30. 31. 35. 56.	1970 0 0	7880 0 0	4538.5$\frac{14}{27}$	18153. 2$\frac{2}{27}$	10800	43200

69 portions of Wood-work as above would cost .	£112,471 10 8
Parapet { For Stockades, 67 × 150 × 2 × 13 = 261,300 cubic feet at 1s. per foot - - -	13,065 0 0
Parapet { For Cross Balks, 67 × 150 × 4 = 40,200 cubic feet, at 1s. per foot - - -	2,310 0 0
Shear Waters, 67 × 40 × 18 = 48,240, at 1s. - -	2,412 0 0
Add for waste of Timber, $\frac{1}{20}$ of the whole (say) -	6,497 18 6$\frac{3}{5}$
Add for Carpenter's work, $\frac{1}{40}$ ditto (say) - - -	3,248 19 3$\frac{1}{5}$
Total Cost of Timber in material and labour	£139,705 8 5$\frac{3}{5}$

247,104.15

Cost of Concrete in the parapet } 23605.0.0

Total cost of Concrete - } £252459.15$\frac{4}{7}$ in material and labour.

There is an error in this left-hand column: the integer and fraction in the shilling column ought to be doubled, and £31. 10s. added to the sum in the right-hand column.

Superficial feet in the vertical face of work . . . } 635250

$\frac{635250}{80} = 7940\frac{5}{8}$ { Cubic ft. of wood in do.

$635350 - 7940\frac{5}{8} = 627309\frac{3}{8}$

Cubic ft. of concrete in the same.

$\frac{627507}{27} = 23233\frac{49}{72}$ { Cub. yds. in do.

	£	s.	d.
$7940\frac{5}{8}$ =	397	0	7$\frac{1}{2}$
$23233\frac{49}{72}$ =	2323	7	2$\frac{8}{9}$
	£2720	7	10$\frac{7}{18}$

Cost additional of every foot extra Width to the Platform of the Quay and Thickness of the Work.

TABLE III.—Showing the Consumption of Wood for each Tunnel as marked, the Cost of each, and the Amount, if so many be needed, of the whole. The Cost for Cement for Massing the Tunnels is not included. It may be estimated at as much more. (Vide Table IV.)

Tunnels.	Soundings.	feet. ft. in.	ft. in.		£	s.	d.
a a	8	108+ 3·0	111·0	at 50*s.* each foot of length	277	10	0
b b	15	122+ 5·4	127·4	„	318	6	8
c c	22	136+ 7·8	143·8	„	359	3	4
d d	25	142+ 8·8	150·8	„	376	13	4
e e	34	160+11·8	171·8	„	429	3	4
f f	39	170+13·4	183·4	„	458	6	8
g g	40	172+13·8	186·8	„	464	3	4
h h	41	124+14·	188·0	„	470	0	0
i i	41	174+14·	188·0	„	470	0	0
k k	41	174+14·	188·0	„	470	0	0
l l	42	176+14·4	193·4	„	475	16	8
m m	43	178+16·8	192·8	„	481	13	4
n n	45	182+15·4	197·4	„	493	6	4
o o	46	184+15·8	199·8	„	499	3	4
p p	45	182+15·4	197·4	„	493	6	4
q q	42	176+14·4	190·4	„	475	16	8
r r	41	174+14·	188·0	„	470	0	0
s s	40	172+13·8	185·8	„	464	3	4
t t	26	140+ 8·4	148·4	„	370	16	8
u u	16	124+ 5·8	129·8	„	324	3	4
					£8,640	12	8

N.B.—A great variety of opinion exists as to the expediency of having tunnels at all. This Table is introduced for the purpose of showing that such a resource is available.

TABLE IV.—Showing the Total Cost of the whole Work, as per Plans, Model, &c.

		£	s.	d.
Cost of Timber, as per Table, No. II.		139,705	8	$2\frac{3}{5}$
Cost of Tunnels, as per Table, No. III.		8,650	12	8
Cost of Concrete for the Tunnels		6,640	12	8
Tar, Cordage, Incidentals, &c., $\frac{1}{80}$ of the Whole (say)		1,624	9	$7\frac{3}{3}$
Cost of Concrete, as per Table, No. II.	£252,459 $15\frac{1}{2}$+31 $39\frac{20}{21}$ =	252,491	15	$4\frac{23}{27}$
		£411,075	18	$4\frac{7}{133}$

Leaving a margin of £88,924 1*s.* $7\frac{128}{133}$*d.* to cover the expenses attending Steaming, Diving, &c.

Tables II., III., IV. are peculiar to the case at Dover in 1845.

COMPUTATION TO SHOW THE AMOUNT OF DEAD-WEIGHT IN A PORTION OF THE PROPOSED WORK.

---

[*] *a b c d e f g h i k a* is a section of the work in 44 feet soundings, being 70 feet from the upper platform of the work to the bottom of the sea, *i h g f* being the sea line at the bottom.

Area of *e g f*	$\frac{140\times70}{2}$ —	4900
Do. *a h g e*	40×70 —	2800
Do. *k h i*	$\frac{50\times17}{2}$ —	425
Do. parapet, *c b e d* .	10×18 —	180

8305 feet superficial in a section of the work as per model.

8305×150 (the length of the model in feet) . . — 1245750 cubic feet in the portion represented by model.
1243750 - 39400 (cubic feet wood in ditto) . . — 1206350 cubic feet of concrete in model.

$\frac{1206350\times125}{112\times20}$ (the weight of a cubic foot of sand — 67309·7 tons of sand } in model.
$\frac{39400\times41\cdot25}{112\times20}$ (the weight of a cubic foot of deal) — 7255·5 tons of wood }

74564·2 tons of dead-weight in model.

But, if the material used be pebble, that which is commonly called shingle, which bears the proportion to sand of 166·5 to 125, and concreted with chalk, (compressed,) which is heavier, then, computing the mass as solid pebble, we have

As 125 : 166·5 :: 67309·7 : 89656·5 and 89656·5+7255·5—96912 tons of dead weight in fifty yards of the work in 44 feet soundings, as represented by the model.

Item.—In the middle mass or portion of the proposed breakwater, there are 34 portions of 50 yards each. Using 30 as a rough factor instead of 34, we have 96912×30 2907360; that is, nearly three millions of tons of dead-weight, in one consolidated mass, to resist the shock of the sea in the weakest part of the Breakwater: whereas, the whole impact of the sea on this mass, supposing it to set in a direction perpendicular to the line of the work, and the water to be impelled against the wall with a velocity equal even to 25 miles an hour, is only 144,552 tons, as opposing the whole surface of this portion; the slope being, as represented in profile in the above diagram.

---

* The Letters refer to the Plan which was a part of and accompaniment of the Model.

But while it is here shewn that this work might have been done for £500,000, there *was* a time when it might have been realized for less than three-fourths of that sum : a notable opportunity was lost of turning an enormous mass of excavation to this profitable use in Dover in the year 1845, when 1,158,696 cubic yards of (shingle-with-chalk) conglomerate was hoven into the Bay of Dover, by the contractor of the excavation, under the authority of the Dover Harbour Commissioners' Engineer, notwithstanding that the Admiralty *once, certainly*, I believe twice, ordered the discontinuance of this wanton act of mischief.

It is to be remembered that this work proposed by me, and avouched to the Admiralty by an Engineer Officer, is a vast Wooden Cradle, constructed in horizontal layers : and it was proposed to fill it with a common concrete. The Wood-Work would have cost about £200,000; the Concrete the rest. Now, when it was determined that there should be an Harbour of Refuge at Dover on the grand scale, it was declared necessary to have a corresponding Basin within the port; and so, 18 acres were bought and appropriated to this purpose. The old York Hotel and a vast quantity of Warehouses and Stores, were bought and pulled down, and the ground on which they stood was excavated to be available as Docks, for refitting disabled shipping, &c. Hence an excavation of 18 acres to a depth of 40 feet being necessary, the sum is—

$$\frac{4840 \times 18 \times 13 \cdot 3}{9} = 1;158;696;$$

i.e.	4840	Square Yards in an Acre.
Multiplied by	18	Number of Acres.
	87,120	Number of Square Yards on the Surface.
Multiplied by	13·3	Depth of Excavation, in Yards.
Gives - -	1,158,696	Cubic Yards. Solid Contents of Excavation.

And what was done with this excavated matter? It was hoven into the Bay—into that bay, concerning which, as has been already intimated, Sir W. Symonds declared that it was injudiciously chosen for a harbour, on account of the tendency of the water there to silt up; and, concerning which expected result, Mr. Calver has favoured us with an estimate of the cost of each year's cleaning out, even when the depth of the deposit is only 6 inches of *silt;* namely £20,000. And here was a quantity of chalk-and-shingle conglomerate,* hoven into the Bay, sufficient to be-spread its proposed area of 500 acres, with a splendid coating to the depth of 18 inches: and so, practically shoaling it to a depth of from 4 feet to a fathom, just at its eddying point!

But it is not only, also, the loss of that valuable material, which would have reduced the cost of the whole great work from £500,000 to £350,000 that is to be grieved over, but, the cost that will have to be undergone to have all this mass of matter dredged up again.

I have often asked, Who is to pay for this? or, to justify this audacious wrong-dealing with the public interests? but I have failed as yet to get an answer. I mean, however, to drag one forth from the proper quarter, somehow: and, in the meantime propose a *4th Rule of Three Sum for Mr. Walker.*—If it cost £20,000 to dredge-up an area of 6 inches of Silt, what will it cost to dredge-up and get-out re-solidified-chalk-and-shingle, of 18 inches' thickness, from a depth varying from 20 to 70 feet? Nothing less than a hundred thousand pounds!! if so little.

And this Stuff, that was thus wantonly converted into mischief, thus defiantly hoven-out into the *shallow* deep—for the Admiralty FORBAD it—and, not at all to float out into the great deep, as they unphysically ventured to suppose, was what was wanted to help to fill the Breakwater Cradle. It was absolutely, *all* upon the railway trucks, and *on* the rail, and from it, tumbled into the Sea!

Well said the Speaker at the Institution of Civil Engineers,

---

* "The whole of the town stands on shingle. It is all a mass of shingle, as hard as rock, by being mixed with so much chalk."—Hon. Admiral Elliott's Dover Evidence, Q. 1941. So that Mr. Walker's conglomerate was as good as *Rock.*

that "all these great works were being executed without any efficient responsible supervision or control; and that not only the public, and the representatives of the nation, but the Government itself, has been utterly in the dark as to proceedings relative to them," and he added, "the time has now arrived when these matters should be brought before the bar of public opinion." Quite fitting, too.

The Cradle required to fill it, 2,300,000 cubic yards of Stuff, and here was, as nearly as could be, half of the required quantity; and instead of being used, the opportunity was, *not lost* but *abused*, and the Bay, injured; £150,000 worth of matter destroyed, and £100,000 of costs to get it up again; sum total, £250,000! at a *low* estimate.

Small wonder that at the Civil Engineers' Institution the speaker said,—"It was evident that no person should be placed upon such Commissions who was pledged to any System whatever, nor should the execution of any work be confided to any member of a commission by whom that work had been recommended. There were many modes of carrying out these views, and if a better and more equitable method was adopted, *the best professional ability* and the greatest executive skill, could be secured, to the manifest advantage of the Country."

But, as one can hardly believe that the Admiralty, once by Captain Washington, and possibly twice, absolutely forbad this thing to be done, and that, notwithstanding, it was allowed to be done by the Contractor with impunity, perhaps we may find some Honourable Member who will be good enough to ask for Mr. Walker's explanation of this curious fact? I *did* hear that, the contract having been made, had the contractor been hindered from heaving the Stuff into the bay, he would have required some £5000 or £6000 compensation; and so, to save this, or the reputation of the rash and heedless person, whoever he was, that agreed to the contract, this splendid mass of material so wonderfully ready, and on the rail, to do the very work wanted, worth, under such circumstances, no penny less than £200,000, was, as completely goods wasted and destroyed and lost to the Country, as if a government store with as much worth of property in it, had been burnt to the ground.

It was alleged that there was no deterioration of the bay, the consequence of these doings. But the shoaling of the bay was a Fact so sensible, that the Packets which were used to wait for the tide, thence forward took a wider offing for anchorage. And is there no account to be taken for these things? Or is it really the case, as proclaimed at the Institution of the Civil Engineers, that it is a principle with the "administrative skill" of this great Country that all these great works shall be "executed without any efficient responsible supervision or control?" Really it would seem so, when so palpable a case of malversation as the foregoing, can be carried on with defiance, and concluded with impunity. Money wasted; a bay spoiled—Per Contra, the privilege of being ruined by an *Eminent* Engineer! Pleasant! Who pays?

Episode Practical.

"To render the opportunities that Nature has afforded available to the greatest extent, the works that have hitherto been erected, ought probably to be removed altogether."

The accompanying Paper, which I take the liberty here to reprint, and think it well to interpose it between the reprint of the Estimate and the reprint of the Pamphlet of 1846, is one altogether remarkable, considering the state of things it reveals, and by and to whom it was written.

It is to be remembered that the management of the whole Coast of the Empire is vested in the Admiralty: and that whatever is done must be sanctioned and "approved" by the Admiralty Engineers. Here are all the Ports of the most important of our Seas, "planned to be accessible at all times of tide," and, of them all, not *one* answers. Manifestly, they are a cluster, a constellation of failure to which Mr. Farrell's General Principle, will be eminently applicable :—" to render the opportunities that Nature has afforded for Harbour Purposes, available to the greatest extent, the works that have hitherto been executed ought probably to be removed, altogether."* Yet, as they have been all done under Admiralty authority, here is another case in which "the administrative skill of the Country" appears but to little advantage. When the Admiralty Officials make light of the Outsiders, the Insiders must have something better to make a reputation for themselves than Admiral Beechey's Memorandum on the Harbours of the Irish Seas. As these Harbours are all of miserable inefficiency, it will be evident that *the* engineering authority is not, as yet, publicly known in the Irish Sea, or, by those who sanctioned the expenditure in all these miserable cases.

" *Memorandum by Captain* F. W. BEECHEY, R.N., F.R.S., *on the Artificial Harbours of the Irish Sea.*

" SIR,

With reference to the Artificial Harbours in the Irish Sea, I send you a copy of the Memorandum which I have had by me some time.

" In the space occupied by my survey, that is, between Lough Foyle, Liverpool, and Waterford, there are nine artificial ports, which were planned to be accessible at all times of tide.† Of these, six have been erected at the expense of Government, and three by private companies or individuals; viz., Donaghadee, Port Patrick, Howth, Kingstown, Holyhead, and Dunmore, on the part of the Government; and Port Rush, Whitehaven, and Ardglass, on behalf of others.

---

* Note, p. 40.

† The reader is requested to observe that " the Highest Engineering Authority" accepted all these cases, as "*planned to be accessible at all times of tide,*" and, of course, appended to each tne Admiralty " approved."

"Of these nine harbours it is grievous to think that the small harbour of Holyhead alone is the only one at the present day in which, during stormy weather with the wind on shore, vessels can lie in such security* in all parts of its waters; or rather it is the only port which has answered its intended purpose. Either they have filled up and become inaccessible; or their anchorage is so exposed from the erroneous position of the entrance, that they are unavailable and unsafe asylums: and besides, the sites of most of them have been badly selected.

"Beginning with Donaghadee, the first of the Government ports in the north, we find it erected upon a nest of rocks; in the centre of the port, small as it is, there are large stones with only five feet water upon them, just 230 feet within the pier head; and outside the pier head, at one hundred yards, immediately off the entrance, at 240 yards, there is a rock with only four feet water on it; and in the same direction, nearly at 400 yards' distance, there is another with only six feet water on it. The entrance, although 150 feet in width, admits the swell so freely, that with the wind any way on shore from north or south, it is either inconvenient or positively unsafe† to lie alongside the pier, according to the strength of the wind and sea in the offing: so that I have been obliged to quit this harbour in a small steamer on the approach of a gale.‡ Owing to the position of the entrance in this small port, there is neither smooth water, nor room for a vessel to round in, in bad weather, without endangering those lying at the pier.

"Nearly opposite Donaghadee is Port Patrick—a nook with only three feet water right across the harbour (has cost £176,000, requires £35,000 more to complete it), at about a vessel's length within the horn of the pier. It is constructed on a bold rocky coast with deep water close outside the pier heads, and is exposed to such a heavy sea in stormy weather, that the sea, running up

---

* If this was the case, it is so no longer. See page 87.

† A "wrecking pool."

‡ A curious harbour of Refuge!

the tower of the lighthouse and rolling over the pier, is the common sight of the country to strangers who visit it in such weather; and nothing but the consummate skill and nerve of the pilots of the place could so long have enabled the packets to use this port with so few casualties. The berth for the packets, two in number, has been partly dug out of the shore; the vessels, when there is water sufficient, are turned into this dock by a rope cast from the bow to a rock; and, when in, they just fit the ditch.* When there is not a sufficient depth of water, they run a-ground, and bump upon the sand (for which they are provided with suitable bilge pieces) until there is, and then they are twisted round the rock to their berths.

"A vast deal of money has been expended upon this port: the northern pier has been frequently washed down, nearly as fast as it has been pushed out, and the southern one not long ago was nearly washed away, lighthouse and all.† In short, it has hitherto presented a series of struggles between the skill of the engineer and the violence of the ocean; and we might almost say, in which, if the former prevail, he will succeed in constructing a very inefficient harbour.

"Next in order is Howth, a small harbour built at the mouth of an estuary, which had long been throwing out its sandbanks, with the entrance of the port inclined towards the ebb, and, of necessity, soliciting the sand and silt, with which it is charged, to seek its asylum;‡ so that, its having filled up and been abandoned, need not excite our surprise. Outside this port, at 100 yards off its entrance, there is a rock with only nine feet water over it.

"Kingstown harbour, in point of size, is one of the most splendid artificial ports in the United Kingdom, but owing to the width and the erroneous position of its entrance, its anchorage is so exposed that in easterly gales vessels cannot hold on,§ and many which have sought an asylum within the piers have been driven

---

* "Eminent engineers!" † See note. ‡ Page 86. § "Dangerous decoy."

on shore and some wrecked.* *Various schemes for protecting the entrance* have been devised, but none of them have as yet been adopted. It unfortunately happens, that the only side of the port which is safe in easterly gales cannot be appropriated to those vessels most requiring shelter; so that, *altogether, this harbour has disappointed expectation.*†

"Holyhead is a safe little harbour, in which vessels drawing from nine to ten feet water may always lie afloat at the outer half of the pier, and the harbour is so protected that those vessels which ground upon the mud are in no danger in any weather. [The entrance is so exposed, that nothing is safe before it.] In heavy westerly gales alone is there any difficulty in getting into this port, and that might have been obviated in the commencement; at which time also, had this port been wisely considered, a much greater area might have been enclosed without a much larger expenditure. [Admiral Beechey evidently did not anticipate that Mr. Rendell was to work up a million and a half sterling out of this modest recommendation.]

"Lastly, we come to Dunmore, a small port built for the accommodation of the Waterford packets. This port, like Howth, has greatly silted up, and been abandoned by all but the fishing smacks, to whom it is still very useful. [1846.]

"Such is the account of the ports which have been erected by Government. Of those belonging to individuals, Port Rush in the north is so badly placed and ill-constructed that in westerly gales there is less danger at sea than alongside its pier.‡ The run of the sea is so great that scarcely any ropes will hold a vessel, and if she parts she is, in an instant, on the rocks. The harbour has

---

* "Wrecking pool."

† Captain Vetch, however, admires it immensely. He says "it is the most successful work for the shelter of shipping that he knows." It is to be understood that the italics which are in this reprint of Admiral Beechey's memorandum are not the emphasis of the writer.

‡ This *is* a clever thing.

also silted up, with the exception of a small canal alongside the pier.

"Whitehaven presents a succession of artificial harbours, one within the other. The outer one, which is the last built, is a splendid fabric, and encloses a large space, of which the only regret is that it cannot be *made use of in bad weather*,* except as a passage to the inner ports; and that only after half tide. The outer part of this harbour, when first built, had sufficient water for steamers to gain the pier at all times of tide, but now there is only a small ditch, which has apparently been kept open by the action of the paddles of the steamers which occasionally touch there. In this ditch a small steamer might lie afloat if it were safe to do so, *but the swell rolls in* so heavily with strong winds on shore that it would not be safe to be caught there.† The rest of the port has so silted up that it is for the most part dry at low water, and there is only three feet across the entrance from pier to pier. It might not have been possible to prevent the silting up of this port without the use of dredges, but its area might have been better protected.

"Lastly, Ardglass is another instance of a great mistake in point of site for a large outlay. Here was a very limited area shut in by a short pier jutting out on a bold rocky coast, so exposed that the first easterly gale which beat with violence upon its shores washed away the new pier, lighthouse and all, before even the light was exhibited, and the remains now lie in deep water.‡ It may be questioned whether the area of the port, in which only one steamer could lie afloat, enclosed by this pier, if it had stood, was sufficient to justify the outlay.

---

* Wrecking pool.

† This is still another character of Harbour of Refuge. This is not a Wrecking Pool, but a Wrecking *Ditch*. What clever device for wrecking shall we next have "approved" by the "highest engineering authority in the country?"

‡ Here was a glorious example of the advantages of "walls of masonry," where nothing but "the continuity of balks of timber bolted together" *could hold* against the fury of the sea.

"Such is, I believe, a faithful and unbiassed account of the artificial ports within the limits of my survey; and which have cost the country much money. It is true that they are some of the earliest artificial harbours that were constructed by us, and they were planned at a time when steam was little thought of; our surveys of the coast were also very imperfect, and it would be unjust to impute any blame to the engineers, who, as regards the masonry, have, I believe, eminently* performed their work.† But there is much reason for believing that, were these plans and sites to be now submitted to a body of practical and experienced men, scarcely would one of them be adopted.‡

"I am, &c.,
"F. W. BEECHEY,
"Captain."

"*To Captain Beaufort, R.N.*"

The closing paragraph of Admiral Beechey's Memorandum is, unfortunately, not inapplicable to times later than those to which that intelligent officer so leniently points. If the case, *then*, presented an appearance of 1 success out of 9 examples, giving 8 failures to one partial success, what is the improvement in the present time of which we have now to boast? where is even this modicum of success? Is the instance at Dover, or at the Tyne? Is it at Sunderland? or Whitby? or Burlington? or Stockton? Is it even at Holyhead? There, last summer, a friend of mine, who has one of the most perfect yachts in England, strong as skill and means can make her, with a bow as sharp as a knife, and "found" to excess of perfection, found himself, by accident, in that celebrated

---

* It is gratifying to find we can have always eminence, somehow. We find it in this passage judiciously transferred from the Engineers to the Masons. Eminent Masons!

† Wonderful means and no wonderful end. The wonder would have been had it proved any otherwise.

‡ Not so; the sites were desirable: and, by a proper treatment they may *all* be made available. Captain Warren has shown *the* treatment, for Kingstown; and that is the Principle of Treatment for them all.

improved port, and all but parted from his anchor in the harbour. While as to Kingstown, about which such wonderful praises have been sung, it would appear to be so choice a wrecking spot, that the syrens of old would have selected it, if they could, for their most delectable treats in destruction of ships and the wailings of their crews. As viewed, indeed, through the roseate lens of the Admiralty Engineer "in charge of the Harbour Department," this is a very charming place; nothing at all to be desired about it, that it has not.

622. "Dublin Bay," says his interrogator, "is of itself a very fine harbour of refuge by nature?" "Kingstown," answers Capt. Vetch, "is the most successful work for the shelter of shipping that I know." With Kingstown then as the index of Capt. Vetch's perfection, I fear that the Admiralty doings under his inspiration will not tend much to raise the character of our national works. "For here," say certain officers commanding Steamers, of a Brig Copernicus, in a Memorial to the Commissioners of Public Works, Dublin, 1844, "the Brig had two anchors down, which would not hold her; and, in a harbour of refuge, became a total wreck, with five or six others—and they are not the first by many that have been wrecked in this harbour." And the memorialists added, "With so many proofs of the insecurity of this harbour, it is absolutely necessary that something should be done to make it in reality what it is now in name—a harbour of refuge." It is also of this Harbour so highly approved, this most successful work for the shelter of shipping that Captain Vetch knows of, that the Harbour Master writes—"it would be no easy matter to secure ships to moorings, when adrift in such a disturbed sea as is presented in this harbour" [*when adrift in such a disturbed sea as is presented in this harbour*] "in two-thirds of its space, with a violent easterly gale." It is of the same *Asylum* Harbour that he (Lieut. Hutchinson, R.N.,) says—"The heavy ground swell and sea which entered the harbour on the north side, made it a dangerous roadstead. I can designate the old harbour as nothing else, when the tide was in,

than a tub of water set in motion." "All the hawsers I had, were furnished to them,"—the five wrecked ships, viz., the brigs *Hemer*, *Dane*, and *Mary*, and the schooners *Bettys* and *Agnes*; "but, three times the number would not have sufficed to hold them, the rolling swell was so great." A pleasant Harbour of *Refuge*. Capt. Vetch's "most successful work." Lieut. Hutchinson further says, "There was a great error committed in ever facing the inside of the walls of both Piers.* the ragged and rough wall cut up the swell, which the smooth face permits to roll, in easterly gales; and with north gales, causes such agitation on the south side, the sea being pent up."

In short, it appears that this famous and supposed "excellent harbour" is duly furnished with all the modern appliances of "The Art of Wrecking made easy." Dover has been greatly improved in this way, with the excellent assistance of that "eminent Engineer," Mr. James Walker, C.E.—See Lieutenant Worthington's Evidence, questions 1825, 6, 7, already referred to.

The Civil Engineers' Institute speaker said, that "the case of Dover was not creditable to the Administrative Skill of the Country." It would be highly satisfactory to have those instances pointed out, where anything is done, so as to reflect credit upon us. We have store and enough, of one-mouthed harbours and "walls of masonry." Dover, Whitby, Sunderland, Kingstown, Whitehaven, &c., it is therefore time to try for something else: a better plan both of treatment and of building; some of Capt. Vetch's long-ago-imagined "new plans and contrivances." I beg to modify the learned gentleman's suggestion. Not new *Plans*—the *Old* Plan. The *Old* Plan of *Treatment*, NATURE'S OWN *very old* Plan, an Island Screen, with "New Contrivance of Building,"† even that of the gentleman who is "no authority at the Admiralty."

---

* "Eminent Engineers!"

† Paper No. 5, The Sand Trap, &c.

No. 8.

The Pamphlet of '46.

"OLD WINE."

THIS is, in part, a Reprint of the Pamphlet to which I have alluded; *i. e.*, about so much of it as is necessary to shew that, what has occurred, I foresaw; and that the Commission was warned of the consequences, financial and physical, and as to the wasting away of time, which would result if they determined to proceed on their rash and wasteful work in Dover Bay.

Now, as all the existing facts of the case are the falsification of all the promises that were made, I argue, that those who made them, *or*, induced others to make them, either, did not know their business, or intentionally deceived their employers or the public. In either case, they must be held to be entirely unworthy of all future confidence; and it is to be hoped that such Incompetents, or worse, may be no longer allowed to abuse the opportunity, or dissipate the means, that an indulgent, because inefficient, Masterhood—viz., the Lords of the Admiralty, and which, no disparagement to their Lordships, a politically-appointed body must always be—are obliged to commit to such Stewardship.

I take leave, most respectfully, to commend my Lords Commissioners of the Admiralty, to give heed to the subjects thus introduced to their notice, and to study them, by the light afforded by these pages and all other sources that may be available to them. The wise doings of the Admiralty Advisers are already, we find, paraded at the "Bar of Public Opinion," not only in Parliament, but elsewhere; and it is highly disadvantageous that the Admiralty should become unpopular through the faults or follies of their supposed subordinates. Unless, therefore the Admiralty means to become morally, as well as officially, responsible for all that is done and all that is OBSTRUCTED, they will gladly show themselves thankful for having their attention called to these important matters, and in being made by such circumstances to learn their business as Guardians of the Public Interests in this behalf, viz., WRECK-WORK, and ANTI-WRECK-WORK.

I beg leave here, however, to separate the late First Lord of the Admiralty from all who have gone before him. Sir John Pakington has dared to act for himself, without being held in the trammels of the professional Advisers of the Admiralty. He has granted, what I have sought in vain for 14 years from every preceding First Lord, viz., to have these statements and plans investigated by a Commission, and ordered it to be done.

Besides being a practical sailor, which certainly a First Lord ought to be, I feel the present First Lord to be a grand Admiralty Reformer—and the Sailor's Good Friend, to whichever faculty he belongs, whether the Royal or Mercantile Navy;—a character, to whom the whole Country may well make an Ovation.

The history of the appearance of the Pamphlet is as follows:—

I conceived a plan of building Sea Wall in any depth of water, combining ease of construction with economy as to money and dispatch as to time; and I was recommended to communicate the same to the Admiralty; which I also did. I was, thereupon, informed that, at the proper time, the Harbour of Refuge Commission would examine it. The time went by, *and the promise was not kept;* and Admiral, then Captain, Sir George Back, who saw the Model at Dover, told me that there was not a minute to lose, as the Report was completed and about to come out: he advised me, therefore, to go to town immediately, which I accordingly did; and the result was, that the Earl of Haddington, then First Lord, commanded the Commissioners to see me. All that resulted from this was, (for the Commission had incapacitated itself to entertain my plan, having bound and rivetted itself with a resolution, viz., to have nothing but Walls of Masonry; which resolution they were unable to burst, when they discovered that the foundation could not carry them,) the putting on record the fact just last-mentioned, before I called the attention of the reader to the Episode Practical—as elicited from me by queries put by Captain Washington. To the Appendix-Report of that Commission, I refer those who are curious, to know by what means the certainty is established that the Ballast hoven into the Bay at the Beach at Dover, is, *every atom of it*, THERE NOW, and that it has *not* been dispersed in the Channel.

The Model was subsequently seen by Admiral my Lord Colchester, at the instance of the Member for my University, the late Mr. Law; and, on Admiral Lord Colchester's Report to the noble Earl that it was worth attention, by my Lord Ellenborough, at that time First Lord of the Admiralty. His Lordship believed, whatever might be the result at Dover, that such a work might be valuable to fix the Sands in the Hooghly. He desired me, however, to address myself to Mr. Corry, Secretary of the Finance Department of the Admiralty; to whom, therefore, I sent the following letter.

THE REPRINT.

OXFORD AND CAMBRIDGE CLUB, *February* 6, 1846.

[15]* SIR,—I have the honor to intimate to you, that I am instructed by the First Lord of the Admiralty to submit for your consideration, the fact of my having proposed a mode of Construction of Sea-Wall, which combines economy of means and time with facility in building: a mode which, if brought into use, in the case of the work proposed for Dover Bay, will reduce the cost of it from £5,000,000 to much less than £500,000.

* The numbers thus enclosed within Brackets, refer to the pages of the original Pamphlet, and are inserted to enable those who may have a copy of that publication to refer to the statements made, and to compare the reprint with the text.

I do not know, Sir, whether on this occasion, it is admissible for me to go into the merits of the plan, but I would state, as a matter of Finance, that, for the alleged cost of construction of the sea wall, as proposed by the Commissioners, the work at Dover and the three other works would be accomplished, if my mode of construction were adopted: and I do not hesitate to say now, with reference to finance, that regard being had to the nature of the bottom in the locality, a wall of cubical *cemented* masonry will *never* stand at Dover, and that no engineer will be able to guarantee that it will: and that, therefore, in building, or attempting to build a solid stone work at Dover, on the site proposed, you are accomplishing nothing, but the preparing for an infinite outlay of means, to patch up, continually, fissures in the work; against the forming of which fissures,* no length of time will be any guarantee.

That this opinion will be combated by those who have sent in plans, I am well aware: and, when it is considered that those who have done so, stand at the head of their profession, I feel that it is to stand at great odds to assert thus against them.† Nevertheless, I do not fear to do so, and I feel that I stand upon this sure ground in doing so, viz., that it is a fact, that all their proposals to build solid hard masonry in masses, are made on the *surmise* that the bottom of Dover Bay is chalk-rock: whereas, it is, in reality, not rock at all, but chalk-clob, and what is below that nobody can tell; for there has, as yet, that I have ever heard of or seen, been no offer made to *probe* the bottom. For, the only experiments which have been made in searching the substratum of the bay, are those of Captain Washington with his

---

* See note in page 94.

† Had I been aware of the contents of that very curious document which came into public existence about this time, viz., the letter or memorandum from the late Admiral Beechey to the then Hydrographer, already here printed, I should have hardly spoken so deferentially concerning "those who stand at the head of their profession."

anchor, as mentioned in page 216 of the Harbour of Refuge Commissioners' Report, when six out of eight experiments proved that the ground was variously "chalky clay," "chalk," and "soft chalk;" the first experiment giving "chalk" as its result [17],—the second, third, and eighth, "chalky clay,"—the fourth and seventh, "soft chalk,"—the fifth and sixth experiments being failures. This being the case, and as there is no knowing, to any certainty, to what depth the *clob* goes; at what point the *hard* chalk begins; or to what degree that rock surface angulates; and that, if it does so *at all*, and any work of a solid *compacted* form be placed upon it, there will be a settlement of the work from *all* such places; and, as every one fissure causes, sensibly or insensibly, two others, and therefore, every settlement is the cause of a division of the work into two fragmental masses;* the consequence is, that no guarantee can be made, except that of rash assertion, that the work will not fall to pieces in the manner above described.

Nor is it to be forgotten, Sir, that there is another contingency, by no means an improbable one, to be guarded against. Who can say that chalk, *under* the sea, does not vary as much in its natural formation, as the chalk-rock *above* the sea does? Recent experience sets before us some somewhat fearful results, in miscalculation on such a point. A similar miscalculation with respect to the foundation of the proposed sea wall in Dover Bay, would be so much worse than was that in the Tunnel, inasmuch as it would lead to irremediable failure. But the work being once begun, National Honour, as well as National Shame, would compel us to keep everlastingly doing something to it, and at it: so that this work, which is, by hypothesis, one which is never to require any

---

* "Secondly,—That though an upright-wall breakwater (like that at Dover), is correct in principle, its construction involves an enormous outlay; in addition to which, *elements of weakness* might subsequently be found to exist, the presence of which would sooner or later cause its destruction."—[Mr Calver's declaration in 1858, p. 30, and Mr. Calver is a great advocate for the upright-wall breakwater.]

repair, would become a yawning gulf into which, half the available revenue of the Country might be poured without producing any satisfactory result.* [18]

Hence, Sir, it is my conviction, and that also of many beside myself whom I could name, that, as a matter of Finance, if the figures of the two plans were reversed, and that my plan were to cost £5,000,000, and the Commissioners' plan but the tenth of the sum, even then, my plan should be adopted, as being the better of the two; better, in as much as it is the best adapted, not only to the circumstances of the case, as they at present present themselves, but as to what they may be, and most probably are. And while I say this, with respect to Dover Harbour, I would, at the same time say, also, very respectfully, that *whatever* Site be fixed upon, no general rule can be laid down as fixing invariably the material of which any proposed work *must* be constructed; for the nature of the superstructure must be made to assimilate, in some degree, to the bottom on which it is proposed to place it. Rock bottom (and gravel) may have solid masonry, or solid *hardened* concrete, as its superstructure; a soft bottom should [19] have a mass placed upon it, such as will sink into the cavities formed by the squeezing down, or oozing out, of the substratum on which the mass is bedded, and which, *i. e.*, the superincumbent mass, may re-adapt itself, in renewed form to its new position. Lastly, a sand bottom must have sand superstructure. If this principle be carried out, you will build cheaply, because, securely; not so much for little cost as upon the best plan; for the best economy, doubtless, is, not to *save money*, but to do, what you have to do, well. But as I believe, Sir, that both of these things are combined in the plan I have offered to the consideration of the Government; and that economy, in its best sense, in the spirit rather than in the letter, would be secured in its adoption, I take

* The proof of this will appear hereafter, in the Dover Engineer's Quarterly Reports. *See* p. 99.

courage to urge it on the attention of those in authority. And as it has received, and continues to receive, the concurring verdict of so many, and those so well qualified to judge of the merits of the question generally, no less than of the plan in particular, I confidently submit it to the judgment of any tribunal that the First Lord of the Admiralty may be pleased to appoint; venturing to suggest that, while there is nobody whose unbiassed judgment I should hesitate to engage with, there are certain parties of high professional name whom it would be convenient should *not* be consulted in the matter: and if I be asked who these persons are I will give their names, and assign the reason why I think as I do.

I have the honour, &c.

F. A. GLOVER, M.A.

The Right Hon. H. T. L. CORRY.

P. 20.

This last appeal to the authorities producing no answer but a bare acknowledgement, I was tired of further endeavouring to obtain a hearing; so thought it best to set the case before the Houses of Parliament and the Public, in a pamphlet which was (not sold, but) sent, in one Hon. Member's name, with his kind permission, to the Port and Sea-board Members of one line of politics; and, in that of another, to those of the other party. I then wrote—

[25] "I have now done with personal and private application. The *Public* may see to its own case. If the Public think it is a pleasant thing to pay taxes to throw them into the sea—if the public think that the lives of our hardy seamen, the property of the subject, the prosperity of the nation, should be equally disregarded—that the Report of a Commission as to *material* may be maintained, to the frustration of its own desire of permanency, be it so! But I can hardly suppose that of Yarmouth, Cromer, the Tees, Arbroath, Widmouth Bay, &c., will be allowed for ten

years to be without those works, the possibility of constructing which the mode recommended by me exhibits, for any so entirely inadequate cause: works, which are quite as much needed *to day*,* for the safety of Shipping and Men's lives, as the proposed refuge harbour at Dover is: and this more particularly, when the sole cause, in reality, why their and the public interests are to be neglected, and why the money, means, and time, which should be devoted to them, and would suffice for the accomplishment of sufficient works for all these places, is, that a *permanent* † work may be erected at Dover.

[26] "This observation leads me to note upon the foundation in Dover Bay, on which it is intended to construct the sea wall, and to carry out the proposition of the Harbour of Refuge Commission. The recommendation of the Commissioners was, that, for the construction of breakwaters, "and for the security of works of defence upon them, the erection of walls of masonry" should be resorted to. The consequence, therefore, of their laying down this general rule, is, that a recommendation in favour of any other mode of building than with "walls of masonry," is held to be a practical opposition to the Commission. This then will account, in a great measure, for the official opposition which has been practically offered [27]—in the *mental* resistance of all who feel themselves bound to support the Commission,—to my proposition.

---

* The year of publication, viz. 1846.

† "The Commission prefer, for the construction of breakwaters, and for the security of the works of defence upon them, the erection of works of masonry." P. 67. Captain Washington told the Commission, that this recommendation applied only to Dover, Portland, and, I think, Harwich; but there is no mention of such qualification in the recommendation; which indeed proved to be a determination rather than a recommendation. Hence, the writer of this was meant to have been morally crushed to death, by Sir Byam Martin's asking him if he meant to recommend wood for material, when the Commission had pronounced for "walls of masonry."

"But it is, truly, to take quite a wrong view of my effort, thus to suppose. It ought, rather, to be considered as aiding to give effect to the general principle of PERMANENCY, so properly held by the Commissioners; for, while I believe that the MODE of CONSTRUCTION which I propose, is a real means of securing the accomplishment of their own laudable desire, I am equally sure, that any plan which they have been considering, if formed upon the idea of solid stone structure being carried from the present bottom of the sea upwards, *must* fail, if attempted to be carried into effect.* *I repeat*, it is not possible to build cemented masonry without a *sound* foundation; and there is none *to rely on* in Dover Bay. It is true, that, when they shall have excavated, if they can do so, or squeezed out the *soft* chalk which lies along the bottom, to the depth of *x* feet, they may come to chalk-rock, and probably will. But when they find it, what is the nature of "chalk-rock?" Let Mr. Cubitt be asked what was *his* experience of chalk-rock in the Shakspeare Cliff. Let the Company be asked, who have had the work of finding money for paying for the making and lining with brick, &c., two complete tunnels instead of one, *because* chalk-rock is *not* to be depended upon for equality either of strength or of consistency. A mistake on this point, jumping at a conclusion and *hoping* that it is a sound one, will not alter the reality of the case, as the South Eastern Railway Company have found. Let England take care how, now, she embarks her £5,000,000 upon a like unfounded supposition, viz.—that because the chalk is hard *here*, it will be hard *there*."

[Nothwithstanding the guarantee of all the eminent Engineers, or of those of the eight who sent in plans for the carrying out of the work, by means of perpendicular walls, in masonry, how many of them foresaw this contingency? The advocates of the Long

* Mr. Calver: see note page 94.

Slope were not called upon to provide against it. There has been, meanwhile, in January, April, July, and October, 1855, rather a melancholy proof of the wisdom of this suggestive caution, and a practical censure on those who despised it, or, thought themselves not bound to listen to the facts of nature, unprofessionally advertised,—that is the way to double the millions of an estimate!—in the Engineers of the Dover Harbour Works' Quarterly Reports—thus:

23, Great George Street, *April* 4, 1855.

Sir,

The progress of the works has been much delayed by the weather; and also from the chalk foundation not proving so good as in the portion near the shore, as stated in our Special Report to you, dated the 29th January last.

Again:

23, Great George Street, *July* 2, 1855.

Sir,

The surface of the chalk being still of inferior quality, it has been necessary to sink the masonry a considerable depth into it, to obtain a good foundation.

Again, in the Quarterly Report, 10th October, 1855.

Sir,

"The foundation courses have been extended 920 feet from the commencement of the pier, being an advance of 28 feet during the past quarter; [this appears to be a splendid progress! but we note that] the progress of this portion of the work has been retarded from the necessity of removing a large quantity of soft material before the foundation could be laid."

We are, &c.,

WALKER, BURGES, AND COOPER.

To the Secretary *of the* Admiralty.

"*The chalk foundation is not proving so good as in the portion near the shore.*"

"*Necessary to sink the masonry a considerable depth.*"—all the stuff to be dug up by the divers.

"*A large quantity of the* SOFT MATERIAL *excavated* [*!*] BEFORE *the foundation could be laid.*"

Was not this what was foretold in the pamphlet in question, written in '46, for which the writer was marked as a malignant? and triumphantly crushed by being reminded that "eminent engineers said so-and-so." "There are 'no engineering difficulties.'" "Have not eminent engineers said it can be done?—and who are you?" I venture now therefore to repeat the question, which I then rather unceremoniously put in the pamphlet of '46, "Who pays for all the broken promises of the 'eminent engineers?'"

In page 55 of that pamphlet, I read thus, "This casual mention of hard substances and soft bottoms, reminds me, before I conclude, of a conversation I had the other day, with a railroad contractor and supervisor [at Dover].

"What do *you* think of the Bottom on which they are going to build their solid wall?"

"Oh! it is good bottom, Sir."

"Good? Suppose they meet with some queer place in it, such as you found in the Tunnel?" (the Shakspeare.)

"Oh Sir, that's only a matter of expense, they must dig it out, that's all."

"Ah!—and suppose they should chance upon some such soft place as Sir Christopher Wren found, when, in sinking for a foundation for that tower of St. Paul's, he had to dig down some 300 feet, until he came to the Paris slab?"

"Well, Sir, they must go into it. That's all."

Go into it! What, Men, or Money, or the Breakwater, or the Commissioners? We have now and thus a foretaste of what "going into it" is.

BUT, THE MONEY! WHERE IS IT TO COME FROM? Perhaps some Member will endeavour to find out this from the Chancellor of the Exchequer:—Money, to enable them to lay a good reliable foundation at the bottom of an excavation some 100 or 150 feet deep, below the level of the bottom of the sea. We shall soon realise, in reverse, the Engineers Institute's £40,000,000, without the aid of Compound Interest, if this befalls. And why shouldn't it? Dover is right in the line between St. Paul's and Paris.]

[28] "It fortunately happens, also, that this Shakspeare Cliff experiment, has shown us, in time, other results. For, not only was it found that the chalk would *not* stand without a sheathing, but another disaster befel the Company in this Double Tunnel. So great was the irregularity of the chalk formation, that in addition to evils arising from the variableness of quality in the chalk, large embedments of sand were found, by some means, to have formed themselves in the chalk mass: so that, in places, masses of the pier, left between the two tunnels, fell away, and thousands upon thousands of bricks were swallowed up, with the due proportion of cement, in rapidly accumulating solid masses of artificial structure, to keep the venerated "Shakspeare" in its integrity above: otherwise, there might have been a series of deep holes here and there, disfiguring that surface on which so many of us have delighted to stroll and look upon with placid eyes.

[29] "Hence, I believe it might almost be said, that, practically, the cost of making and sheathing *three* tunnels has been incurred in boring through the Shakespeare Cliff; owing, in part, to the rottenness and variableness of the chalk; and, in part, to the eccentricity of the geological formation, and because it was *thought* that *solid* chalk was sound; and that, because it was hard *here*, it would be hard *there* also.

"And what is *now* proposed to be done by the engineers? At no greater distance then a mile from this experiment, to risk the lodgment of £5,000,000 worth of compact material and labour, on a chalk *untried* foundation. I maintain it to be a fact, that up to this time, no effort has been made to probe the bottom of the sea, or the intended site of the work. What then must be thought of a recommendation which asserts the propriety of building in a *certain* way, on a given site, when there is no proof of what the foundation is: and when, unless it be rock, such recommendation cannot be carried into effect? It is very proper that commissioners should *recommend*, *permanent* works to be erected; but the mode of accomplishment *must* vary according to circumstances; and whoever determines *in what manner* permanency is to be obtained, should have a perfect knowledge of the realities of every case proposed. Now I venture to ask of [30] the Commissioners, one and all,—all those, that is, who *have voted for vertical walls and against the use of concrete* in that locality—when they asserted that "walls of masonry" should be placed in water 60 feet deep, out in Dover Bay, on a chalk clob bottom, were they aware of the fact I have mentioned as to the nature of the borings in the Shakspeare Tunnel, and which I have, herein, set before them and the public? I am not blaming the Commissioners for not knowing them. Who is to know the soil and substratum of every locality in England? But they will be greatly blameable, in my judgment, if now, that they know what has been seen in the neighbourhood of the place which they have chosen as a site for the proposed work, they venture to adhere to their original dictum. I say, that recent experience should have caused them to pause and to hesitate, ere they continued to take upon THEMSELVES the RESPONSIBILITY of requiring £5,000,000 of the public money to be spent upon what becomes, under such circumstances, a mere crotchet: that crotchet being, that an everlasting work, worthy of England, shall be constructed! I venture to predict that it will be a work altogether and most unworthy of England, if they carry their point, and spend anything less than £20,000,000 upon it; and that then, if they succeed, they will have accomplished nothing [31] more than the building of a splendid bauble: a thing which, in comparison with what might have been constructed for one-tenth, not unlikely *one-fortieth*, of the money, will be, *in every way*, practically inferior to the latter.

"But it will be perhaps said, "Who are you who speak thus? Have not engineers given their opinion, and don't they say it can be done?" As to that, I don't know; [see Paper 1,*] but if they have, there is something in the way in which a man gives his opinion that lets us into the real opinions entertained. One gentleman, for instance, the same who had experience in the tunnel spoken of above, brings forth the project of sinking caissons 50 yards long. Mr. Rendle, on the other hand, in his evidence, (p. 180, of the Report of the Harbour of Refuge Commission), declares against having recourse to any such gigantic efforts of caisson; and says that, in smooth water, in a river, with plenty of assistance, he found caissons 20 yards long almost unmanageable in 27 feet water. What then do such propositions show? This, surely: that there are, or are held to be, objections almost insuperable to things being done in the ordinary sensible way of laying a foundation before you begin to build; and therefore recourse must be had to extraordinary ones.

"But it may be said, and will, perhaps, "You [32] are not a professional man, and these gentlemen, who have propounded plans, have professional reputation to sustain their assertions." That is a great thing for them, but what is it to the Country?†

---

* *See* Mr. Walker's Evidence before the Select Committee, answers to *Q*.3825.

† We can answer this question, if not accurately, yet somewhat *in extenso* now. This job was promised for £2,500,000; it now shews a brazen front of £5,000,000; it won't cost one penny less than £10,000,000, if continued on the present system. Lowestoft cost £200,000; which the Government sold up, and got £4900 for their £50,000 lent; the shareholders, for their £150,000, got 0. Holyhead was *promised* by the late Mr. Rendle for £650,000; it has even now advanced to upwards of £900,000 for the rubble work; the masonry has yet to be begun; it will cost not less than £1,500,000, and lucky if it be so little. When the Hydrographer was asked, "how Mr. Rendle could have ventured upon such an assertion and promise," his answer was "Mr. Rendle was a very sanguine engineer"; and anyhow the goods are worth the money. That, is as people may think, and as the harbour may prove.

[We have seen that the Commissioners recommended that £5,000,000 worth of solid masonry should be laid upon a foundation of which they knew nothing; nor did they make any provision for probing it: and we have seen, *in part*, what has come of this culpable neglect, as shewn by Messrs. Walker and Co.'s Quarterly Reports.

Contrast now the conduct of these Stewards of the Public Purse, with that of those who have to look after their own money. A small Company with a capital of some £25,000, proposed to make a pier, at Anglesea (I believe it was), opposite to the Isle of Wight. How different was their course to that of these magnificent Commissioners. They spent, so I am informed, nearly £2,000 of their capital in assuring themselves of a foundation before they ventured to build. These gentlemen very properly preferred securing a good foundation to "going into" a bad one.

But the most marvellous part of the proceeding was, that, although the Royal Harbours of Refuge Commissioners never condescended to give orders to probe for a foundation, yet, when they had been driven (by the pertinacious opposition of one of their own body who refused to sign the report,* on the ground that the anchorage was bad in Dover Bay,) to try the bottom, and had found it was *soft*, and therefore entirely unfit for, and incapable of carrying, stone work, *they still persisted in recommending* "walls of masonry!" Is it possible to believe in such curious development of advanced intellect? Such are the facts of the case. Reader, judge for yourself. And this was a commission of eminent men. Some are born to greatness, and some achieve

---

* "I dissent from the Report, because I consider the evidence to be in favour of Dungeness; and because I cannot recommend a large close harbour at Dover, where the pilots consider the holding-ground generally indifferent, and the engineers say it will silt up."—Capt. Sir W. Symond R.N., Surveyor of the Navy.

it. This will be the eternal glory of the Harbour of Refuge Commission of 1845-6, "Cemented Masonry on a soft bottom!"

It will be therefore, still, I hope, allowable to ask—

[32] "Will the 'professional reputation' of the gentlemen employed restore the *enormous sums of money*, which, in several recent instances of notorious mistake, have been utterly, irretrievably squandered?" But that is a trifle. Unfortunately this loss of money *is* a trifle to those who make it their business to lose it for others. It would be well, however, if those who are entrusted with its distribution and outlay, were bound, to a certain extent, as securities, for the ultimate success of their plans. For instance, to rise from small men to great, let me ask,—If the Harbour of Refuge Commissioners were bound, to the extent of their private fortunes, as to the soundness of the foundation on which this £5,000,000 worth of cemented "walls of masonry" is to be placed, would they have recommended this work to be of walls of masonry? I don't mean to say that they would not; but I am very sure that, 'ere they did so, they would have thought it well to expend more time in boring into to probe the bottom, than they have in making their report. And if some of the Civil Engineers were bound to repair their own mischief, and to compensate their own [33] waste, when such could be proved to have resulted from ignorance or perverseness"—[*see* Mr. Walker's evidence*],—"though they could not, in every case, make full amends, we should have the satisfaction of feeling that *something* had been paid towards the loss incurred,—the damage sustained. But no—these are *not* the results we should have. Under such an understanding, and such undertaking on the part of Commissioners, Engineers, &c., we should not have men so ready to vote, nor so handy in dealing out, awful sums of money; nor so heedless of reasonable probabilities, as to pooh-pooh every thing, however

* See p. 69, and Dover Evidence, 1836, quoted pp. 114-5.

commended by local experience as adapted to the exigencies of the place, or to common sense by soundness of principle, that does not come to them under the high-sounding sanction of professional reputation. This too may be pooh-pooh'd. However this may be, the remonstrance meant to be expressed above, is, in all respect, intended to call men to their senses; and to invite the Members to challenge themselves, as in the presence of a whole People—a Maritime People, and who are called upon to judge of such rhetoric,—with questions such as these:—

"Shall we lavish £5,000,000 in doing, in one place, in 10 years, what may be done, in ten places for less money, in two?

[34] "Shall we, the necessity and facility of protecting them being proved, leave our ships and men's lives in danger, and sacrifice them, for so long, to a fantastical notion: a notion, absurd, even if the expectation on which it is based, prove correct, but, which becomes wicked if it do not?

"Shall we, with the experience of the soil in the neighbourhood of a given locality, hazard £5,000,000 in building a work, which nothing but a good foundation can make permanent, in preference to throwing down a work that may be constructed for a tenth of the cost, and which will form, at the worst, a solid FOUNDATION and CORE for any work that may be required hereafter? and this, a Work, which will, until then, answer every purpose for which it, or any other work, can be required in such a place? And shall we, with such and so many glaring instances before us, of the inefficiency of 'celebrated civil engineers,' commit our judgment heedlessly to the opinions of such men" [*see* Mr. Walker's evidence], "and shut our eyes to the evidence of experience?—to the nature of the localities?—to the witness, alike, of Common Sense and Nature?

"And lastly, shall we allow the recommendations of a Commission" [*see* Mr. Walker's evidence], "given in general terms, without reference to, and also for, the particular exigencies of localities, to be carried out to our [35] ruin, and to the defeating of the very object they had in view in the laying down their own stipulation?"

[I venture now to call the attention of the Lords of the Admiralty, and Members of Shipwreck Committees, and Refuge Harbour Commissioners, and Members of both Houses, and the Maritime Interest, to these questions, *now*, after this lapse of time, and to ask them—in the slang of the engineering craft, "with every confidence"—for we'll drop the modesty, and only consider the facts—whether or not they were *good* questions? and good *cautions?* and whether or not events have declared and justified the wisdom and foresight that dictated them?]

"But what then,—are we to have our Commission to sit for naught, by asking them to stultify themselves, by a reversal of their own decision? *Parturient montes, &c.?*"

To all this, it was answered by me, p. 40.

"Better that they should deliver themselves of a mouse, than be allowed to devour us with a monster, vast, and mis-shapen.

"There is no occasion for them to forego their Plan, in adopting, *as to Dover*, a Mode of Construction which provides for them a foundation, on which, and by means of which alone, their own suggestion *can* be carried out; a mode, too, which combines the greatest economy with the greatest assurance of success and durability. It surely would redound greatly to the honour of the Commissioners to say, 'Misled by the confidence of their professional adviser,' [who evidently knew nothing about the matter—*see* Mr. Walker's evidence] 'as to a point of building detail, and ignorant of the nature of the ground in the Shakspeare Tunnel and of what had occurred during its excavation;—and also, finding that the experiments in the Anchorage shew a large amount of soft bottom in Dover Bay, of which the depth is doubtful,' [*see* Quart. Statement], 'it was found better, for the purpose of more fully giving effect to their own recommendation of 'permanency of construction,' that means should be taken, in the first place, toward the formation of a sound bottom, to be a foundation for work to be ultimately decided upon; and that they had, therefore, thought it right to determine and advise, that effective means be taken to accomplish such essential preparation; and that, consequently, all work in "walls of masonry" should be postponed, until the due seasoning of [41] the foundation, by its settling down, could be reported upon as having taken place."

[This was *not* done, and we have seen the consequences, in the Quarterly Accounts for the year 1855. *See* p. 99.

Is it to be done with respect to the part of the work that remains to be carried out? or, is the high horse still to be ridden, because "great men" won't be taught? Be it so. Nine millions, Gentlemen of the Commons, is what you will waste of the people's money in this behalf, if you any. longer condone this insolence, that refuses to be taught by the experience of fact.]

[61] It so happens that the material best adapted for the purpose of carrying out the work at Dover, is close at hand; and, happily, abundant, and useable, almost in its natural state. It happens also to be cheap, that is, it is to be had for the digging of it. But if it were a thousand miles off, as it is the best material that could, both *with* shingle and *without* it, be used for [62] the purpose, so would that be money well spent-which went for the paying for fetching it from thence; and it would be cheap for such a purpose at any price. But since it is where it is, the plan accomplished by its means, under the mode of construction which I have suggested, will not cost more than £500,000. The Harbour of Refuge Commissioner's plan is to cost £5,000,000. A wonderful difference! But there is a greater difference still: for, to guard against contingent results, the Engineer* will have *a right* to demand an *unrestricted* recourse to the public purse [*see* p. 36, same pamphlet], *over and above the amount of his estimate.*

---

* Some idea of the value of these "contingent results," may be now formed, though inadequately, from the Engineer's Quarterly Reports, already quoted, p. 99.

"[63] Under the statement of all these circumstances, I cannot doubt but that the object for which I have printed this pamphlet will be accomplished; feeling sure that some member of the House of Commons will move for, and be supported in his appeal to the House, for the appointment of a special committee to examine into the allegations of its pages; and that, this being done, the Admiralty will be relieved from the dilemma in which it was falsely placed, as seeming, on the one hand, if they were to show a disposition to act upon my suggestion, to disparage the recommendation of the Harbour of Refuge Commission, as to material; or, on the other, by maintaining [64] that, to reject a plan which, besides being better in itself, promises collateral advantages, that cannot result from the adoption of the worse. Inquiry and investigation is challenged, that evidence may be taken as to the general fitness and durability of the material proposed for use. It will be for the Guardians of the Public Interest, no less than of the Public Purse, to consider whether the assertions there made, either as to Finance, Material, Foundation, Time and *Places*, are so trifling as to warrant them, unheedingly, to vote away £5,000,000 for doing, or attempting to do in ten years, at a risk, what may be done to advantage in two years, with a better material, and for a tenth of the money: in short, for trying to do one work in one place, while others, ten other places, which cry out on behalf of the drowning seamen, are being and must be, utterly neglected.

[The appeal was made in vain, and the money was voted, but I verily believe that somehow, "not only the public and the representatives of the nation, but also the Government itself, were all utterly in the dark as to these proceedings." The solution of the difficulty is, "the schemes proposed by certain parties were sure, by some 'hocus pocus' to be generally recommended." Refer to Speech, p. 7 and p. 9. The pamphlet concluded :—]

"It is right that I should state that my proposal goes no further than the mere construction of a sea wall, surmounted by a parapet, by an accumulation of material in a particular way; and so, as to furnish a sufficient work of *defence* against storms, with *space* and *fonndation* for works of *offence* against an enemy, if hereafter needed; employing for this purpose £140,000 worth of timber, as a mere mould into which to fill £250,000 worth of concrete, such as I have [65] spoken of; a mould which would last for ever; but of which, the rotting, if it occurred in a week's time, would be of no importance whatever. My estimate does not, therefore, include the cost of making casemates, towers, bastions, or light-houses. The parapet may indeed be embrasured, which I should not recommend; or it may be crested with guns on traversing platforms, which I should. And this arrangement, though but temporary, *if temporary*, would have this great advantage over guns in casemate, that you would be able, at any time, to direct salvos from a hundred guns against any particular ship in the offing;—a favour that you could never dispense from guns in casemate.

"The reader then is requested to understand, that what it is proposed to do, by the mode of construction to which these pages allude, is, to form an HARBOUR of REFUGE merely. It will be time enough to talk of WORKS of DEFENCE, when you have secured a FOUNDATION that will carry them. When you have done this, you may, at discretion, and at leisure, begin to build Stone-Work and "Walls of Masonry," *from just below low-water mark;* and, if needful, which I hold it would not be, to line the whole of the inner face or quay-side, down to the entrances from the Sea with masonry. By waiting, you would then, as regarded the inner face, secure the threefold advantage, of [66] building on a well-compressed bottom, and always in smooth water, and against a well-compacted side; and, as respected the outer face, you would diminish the ultimate expense by getting rid of the cost *and* difficulty of building some forty feet of sub-marine "walls of masonry" in deep water, and in the open sea. Hence, by the adoption of the mode proposed, while you would realize, at once, the advantages of an Harbour of Refuge, of Defence and Offence, you would *secure*, ultimately, as well as all those objects which the Harbour of Refuge Commission have contemplated, in their proposal to carry out the work in question, by walls of masonry, "A Wall, Platform, or Quay, of 92 feet width; affording space, both for houses of merchandize and for all needful military operations; but which, by attempting to accomplish before you make a foundation that can be relied upon to carry your walls and works, you will never realize at all."

Here ends the Pamphlet of 1846.

It is not to be supposed, however, that because the appeal to the Members of the House of Commons was made in vain, that the pamphlet did not produce an effect.

The commencement of the work that was wanted immediately, was postponed for two years, on my account, as I was somewhat acrimoniously informed by one who had an interest in knowing: for so long, indeed, that those who had felt it right to watch the measure had their suspicions altogether lulled by this season of supposed inaction.

It was during this time, that Col. Sykes congratulated me in having "at all events, prevented five millions being thrown into the sea" at Dover; while another gentleman was induced by the same lull to assure me, in the streets at Brussels, that the whole affair was at an end. "Why do you speak with such confidence?" said I. "Because," was the rejoinder, "no Government of any country would dare to lend itself to such a job as that."

Nevertheless the project was under weigh, and, in less than two months from that time, the work was commenced; but, with a change in the dramatis personæ; so that the conduct of the affair *fell* into the hands of the mere Cinque Ports official, viz., Mr. James Walker: it may indeed have been "*unwilling* hands," at all events, there is a sentence in Mr. Walker's evidence (3895,) when that gentleman, as engineer of the works, was twitted by Lord John Hay's question, concerning the enormous time that this miserable abortion, the Admiralty Pier, index to the Dover Harbour of Refuge, was "dragging its *short* length along," which might be worth searching into. "There are things respecting Dover Harbour," said Mr. Walker, "about which information can be much more correctly given, by gentlemen belonging to the Admiralty in this Committee than by myself." Strange, this, considering that Mr. Walker had been a member of the original Harbour of Refuge Commission; and was, later, the Engineer of the Works, which the Commissioners were supposed to have recommended.

However all that may be, the danger of proceeding on a certain pre-determined plan was pointed out—the warning was regarded and disregarded. Of the first, *I* have proof; of the last, the nation has proof. The result has been and is, that a miserable stump has cost £400,000, and twelve years' labour. The warning was defied: the bottom was *not* probed: the bottom *proves* to be bad; and the money is lost: *and*, worst of all, the plan is foregone—for the nation is now told that it is *not* to have its Harbour of Refuge at Dover, at all. I call this a cheat. I don't scruple to call it AN ABOMINABLE CHEAT. The nation asks for its harbour and its money, and a stump is pointed at it!

The reader may now begin to understand the utterance at the Institution of Civil Engineers, sent as a motto to the next paper.

No. 9.

Dover Harbour Past, and what it needs at Present.

"Hence it might be assumed, that Works were authorized, and the money of the Country voted away by the Government, without any idea being given of the time of construction, or of the cost of such works, nor even of their mode of construction. Now mark the result at Dover: about £400,000 has already been expended, and yet it was, at times, nearly"—[the learned speaker should have said here not "nearly," but "quite, *utterly*"]—"impracticable to effect the landing, at low-water, of the passengers from the small steamer arriving from Calais, until the outward-bound steamer had left. This inadequate result, after such expenditure, was not creditable to the administrative skill of the Government of the Country.

"In fine, it appeared that but little hopes could be entertained of more rational proceedings on the part of the Government, so long as Commissions were constituted as at present: because members of acknowledged independence were swamped by other members determined by foregone conclusions; and the schemes proposed by certain parties, were sure, by some 'hocus pocus,' to be generally recommended. It was evident that no person should be placed upon such commissions who was pledged to any system whatever; nor should the execution of any work be confided to any member of a Commission, by whom that work had been recommended."

I SUPPOSE that there are few people who would not "raise up both hands" in favour of the magnificent promise held out in the concluding lines of the Pamphlet of 1846, if they could have it realized, against the miserable result presented by the Dover Harbour Authorities, and the Harbour of Refuge Commissioners, and Admiralty Engineers and between them all.—and exhibited in the stupid thing called, par excellence, the Admiralty Pier at Dover. *I* am not ashamed of what I wrote and promised; neither now, to promise it again. Is the Admiralty proud of their performance? It is to be hoped that they are not, though they *have* allowed it to be palmed upon the Public as theirs. At any rate if *they* are proud of themselves, the Country is ashamed of it. In a full assembly of the Civil Engineers' Institution, it was said, but mildly, "this inadequate result was not creditable to the administrative skill of the Country." What wonder?

Let us now turn to the result of the former engineering at Dover by Telford, Walker, & Co., as concerns the Scouring Toy, &c.

1960. [Sir John Rea Reid.] "If I understand Captain Elliott, his decided opinion is, under all circumstances, and with reference to the progress of the works, that you would make no alteration until the present plan has been tried?"—"Why, I cannot say that that is quite my opinion; because I have already stated, I think, pretty distinctly, that I consider that if the whole of the present plan was completed, the harbour, as far as regards a refuge harbour, would be just as imperfect as it is at this moment."

1969. "You have known Dover harbour for a considerable number of years?"—"Yes, I have?"

1970. "Have there been different alterations made in the works since you were first acquainted with it?"—"Yes."

1971. "In your opinion, then, have the alterations that have taken place been improvements or otherwise?"—"Why, in my opinion, anything in the world but improvements. I should say, they have materially damaged the harbour itself by contracting it, and increasing the swell" [by the upright walls] "so as to render it extremely inconvenient to the vessels that lie there; and I believe the Bar to be fully as troublesome, if not more so, than ever it was before."

1972. "Do you consider, from your past experience, that Dover harbour, as you first knew it, was capable of considerable improvement?"—"I do."

1973. "And these alterations will not effect the improvement that you consider it was capable of?"—"I do not think they will."

1978. "Do you not know that the opinion of nautical men was taken?"—"I cannot conceive that the opinion of any nautical man was ever taken, because I never heard any nautical man in my life that did not seem to form the same opinion of the works at Dover; when I say nautical men, I do not mean fishermen, or men of the place, but practical men."

2001. "I think you stated you heard strong opinions expressed by the people living at Dover as to the probable success of the works;—though varying in opinion, yet a strong opinion in favour of the probable success of the works?"—"I am sorry to say, so far as success is probable, the opinions did not appear to me to vary much. I do not think that I ever heard anybody, who is not immediately connected with the harbour, or with the works going on, that seemed to expect any useful results from what is now done. None of the gentlemen I conversed with down there seemed to expect it; and my own naval friends, whom I have met there, all seemed to have the same opinion."

2002. "Is Mr. Walker a man of judgment in respect to the construction of harbours?"—"Mr. Walker is an engineer of considerable eminence; but, in cases of this sort, I don't know what experience Mr. Walker has had. Mr. Walker seems, certainly, to have very great confidence in the beneficial effects of carrying so large a stream of water from the Head, as now proposed; and he is one of those persons, who, I say, having so much confidence, rather shook the reliance I have in my own opinion. I did not like, when my opinion differed so materially from Mr. Walker's, to give a very decided opinion in opposition to the strong one he was ready to give."—Hon. Admiral Elliot's Evidence, Select Committee on Dover Harbour.

[Here is the opinion of Mr. Walker alluded to:—735. Sir John Rea Reid asks Mr. Walker, "In your own mind, you have no doubt that it will answer the purpose?"—"I have not."]

1824. "Do you recollect the state of Dover harbour many years?"—"Yes, 30 years I have known Dover, and have had a perfect recollection of it."—Evidence of Lt. Worthington, R.N.

1825. "Do you consider what has been done during the last 30 years has improved the harbour?"—"Decidedly not; I do not know of a single work that has been carried on that I can really say has improved the harbour. We are as subject to a Bar now as ever we were, and I consider the natural qualities of the harbour to have been quite destroyed. It was formerly known, even to a proverb, that Dover harbour, as a dry harbour, when once entered, was the safest harbour in England ; but its qualities are now entirely superseded by being surrounded by perpendicular walls."—See also answers to Q. 1851, all strongly in the same sense.

1587. "Is the harbour of Dover, in your opinion, a better harbour now than it was 20 years ago?"—"No, I am decidedly of opinion that it is not, and I give that opinion as a sailor.

1588. "I think it was in 1820 they commenced the plan of running sluices through the south Head, and cut off a considerable portion of the outer harbour for that purpose; which, in my opinion, has almost ruined it as a place of shelter for ships in bad weather."

1589. "Previous to this alteration which you speak of, as the first which was made with which you were acquainted, did vessels lie in safety in the outer harbour?"—"I should say so, certainly."

1590. "One of the objects of this plan was to obtain a dry dock?"—"I cannot say that. I think that if that was the object of the plan, they could not have put the dock in so bad a place as they have put it."

1600. "Will you state to the Committee your opinion of the works now in progress?"—"I certainly have given this plan a great deal of consideration, probably as much as any person in Dover, and it is my firm opinion that it will not answer what is expected from it; indeed, I think it will be a total failure," &c.

1677. "You used the expression that it is the general opinion that the work will not succeed; let me ask whether, at any rate, there is not a great division of opinion on that subject?"—"All that I have ever spoken to about it—and I have spoken to several naval and engineer officers—have entertained very great doubt about it."

1696. "When you say the former plan was a total failure, are you not aware that those sluices were commonly successful, and to a great measure successful until the winter of 1833 and 1834, when a continued south-east wind entirely blocked up the harbour?"—"Certainly not."

1704. "Would the adopting of Mr. Smeaton's plan render Dover Harbour a more accessible and safe harbour in south-west gales than it is at present?"—"I mean to say so, most certainly; at present it is not at all approachable."

1666. "Nobody ever treated on Dover Harbour so well as Mr. Smeaton; and I believe I am correct in saying, that Mr. Rennie was always of opinion that Mr. Smeaton's plan was the best; and I have no hesitation in saying, that all the engineering officers I have spoken to, all the naval officers I have spoken to – and I have spoken to men of high rank in the navy—have always said that his is the best."—Report, 1836; Admiral Boxer's Evidence.

Such is the evidence of three experienced sailors; and their prognostics concerning the success of the former works are but too painfully realized; for which result Mr. Walker cannot be otherwise than, *to the utmost*, accountable; for the last words of his Evidence in the matter, 1836, were—"But I hope that the engineering of this work, as respects designs, is almost done with; and I hope there will be no derangement of its progress." [Answer to Q. 834.] And the result of the more recent engineering, which is more especially the subject of this pamphlet, is equally apparent: viz.—After all the money spent, and time exhausted, the issue of all the work, former and latter, is, that a Packet arrives in the middle of the night from the other coast, and cannot land her passengers or mails! She must run down to Whitehaven!

Such is the Dover *present* "conclusion of the whole matter."

In that Assembly, however, where "the administrative skill of the Country" was disparaged, the speaker said something else:—"The time has now arrived, when these matters should be brought to the bar of Public Opinion." It is for the purpose of facilitating this most desirable issue, that, in publishing this book, I have here reprinted the pamphlet of '46, affixing carefully the dates. It boots not now to ask, "Who said?" or "Who speaks?" or "What is he? or, was he?" The thing to ask is, What did he say? What are the *dates?* It is these dates that enable me to appeal with an assured knowledge of the verdict, to that "Bar of Public Opinion."

When I appealed before, to the Public, to the Authorities, and the House, I was, doubtless as one that seemed to rave, and the unthinking among them might, perhaps, think themselves justified in saying, "What boots it, what this mad Master of Arts says?"

After the miserable spectacle that our "Administrative Skill" have contrived to present to the world, in the abortive affair of Dover,—after the enormous waste of money in that place,—and the proof, now patent, that the "Administrative Skill" did not know what was about to be undertaken;—as is also practically shewn, by their being astonished at finding themselves so soon met by a soft hole;—after the disagreement of the promises of cost with the painful reality;—and after my printed declarations in this behalf all having been so scornfully neglected, yet so painfully realized—they will hardly *now* "toss the head," or, for the future, dare to set up their authority as Engineers or Estimators, against mine. Whatever the value of *my* authority, theirs is, evidently, worth not a rush.

Again, also, they no more knew what would be the Physical Result of running out that Horn to the sea, against the bay shore, than they did (as now appears by Mr. Walker's evidence,) how they were going to set about the work, how long it would take, or what it would cost. And if this was all done under "the *best* engineering authority in the Country," and I was able to point out how and wherein they would fail, and *they have failed* in all these particulars, besides in that of having thrown a barrier right across a tide way, without leaving passage under-sea for the tide stream;—then, "the best engineering talent in the Country," which they claim to have, to be, and to represent and command, must be content to take a lower stand, as such, than that that is the due of the man who points out their faults, and predicts those results, which they could not see for themselves, and were unable to estimate properly, even when they were set before them. Hence it is, that, when such persons, Corporations-sole or otherwise, were deriding me and my sayings and printing, little they thought that they were laying the basis for conferring on the object of their contempt a Diploma to take precedence of themselves.

When, then, in exercise of the privilege of this Diploma, I ventured, when I was asked by the Royal Commission, in December last, "What you should recommend as to the works at the Tyne?" to answer, "Certainly, not another stone should be laid", so do I say the same, with respect to the Abortion at Dover. The farther they go out with that stupid projection, the more they endanger the Esplanade and Crescent. They will do nothing by turning the corner, but make it more difficult of entrance; and, as to supposing that they will be able to divert the shingle from entering the harbour when the new foreshore is formed up to the mouth of it, they may as well expect Newton's Apple to rebound from the ground, and re-establish itself on the tree: while, their Inner Quay, upon which they reckon to quay the Packets, will all become a shelving beach within. As much as possible they should revert to the Plan of Mr. Cubitt; and as to Mode of Construction do, at last, what they ought to have done at first, use the plan recommended by Professional Skill and Experience, and Common Sense, viz., one of those new "Plans and Contrivances" after which the Admiralty R. E. Official has been seeking for 20 years, and give up the nonsense of Building for Eternity in "walls of masonry":—in one line, recur to Mr. Cubitt's Pan of Harbour, and use Mr. Glover's mode of Building. Such should be "the Dover conclusion of the whole matter."

If a man come forward to propound anything not apparently

in his own line—though I must confess, I know nothing within the whole range of mechanics or physics, which should be foreign to a Master of Arts of the University of Cambridge, unworthy as I may be of that honourable distinction—he is looked upon with suspicion ; not so much by the Outsiders generally, as by those of the outsiders who are influenced by the small talk of the *Insiders ; i.e.*, such as feel that " the craft is in danger " when those not affiliated give their opinion, either as Interlopers or Reformers.

It is unfortunate, either for the Public or myself, that I cannot write R.E. or C.E. after my name.—Had this been the case, 14 years ago would the Harbours of England have been begun to be treated, and the Island Theory have now been the Universal Rule of Treatment. I might, indeed, if I chose, stand upon the fact of having been appointed Colonial Engineer at Sierra Leone, in the year 1825 ; but why be beholden to such or any other authorization for license to speak or advise, on a matter which commends itself to every man's common sense. Is it not enough that men can see for themselves—1. that an Island is a protection to what is anchored behind it ; and, 2. that a Harbour with two Entrances is better than a Harbour with one Entrance ? Must a man be a Civil, or Colonial, or Military Engineer to enable him to see this, or that he may be allowed to say it ? Surely any body may see and say this without Special License !

As however I was bold, in an early page of this tract, to claim for myself a good degree in the Noble Science of Harbour Making, I have felt it proper, in order to make good and substantiate that boast—for it IS A BOAST—to shew cause for having some right to have this claim recognized. And if the intelligence on the subject which dictated the papers, One to Seven, in which is laid down, maintained, and well supported by various authority, the Great Principle of the Island Harbour, together with a Principle of Building which shall make it universally applicable ; and, collateral thereto, a system of Sand-Trap, by which Sands may be fixed, and worked up into Islands, and Estuaries changed from Wrecking Grounds into the Scavengers of our Shoaling and ever-shoaling Seas and Channels—doctrines by me maintained for 14 years without intermission, and now receiving still and daily, the ever-accruing testimonies of those qualified to speak, as extracted from the evidence of the Parliamentary Blue Books, I say, if such intelligence on the subject be not held sufficient to authorize me to speak as a Doctor on this behalf, even without Diploma, then I claim the right to do so, in the fact, that, 13 years ago, the Pamphlet was written and printed, in which is pointed out that that escaped the observation of all the "eminent engineers" in the Country, and which, nevertheless, ought to have been pointed out and WAS NOT; and also, that I knew what the cost of work would *prove to be*, which, those whose business it was to know, either were ignorant of, or held back from declaring. Therefore, if Knowledge constitutes the right to speak and advise in the presence of proved and mischievous Ignorance, I have, by the pages of that Pamphlet, established my claim and my right, in the presence of all who ventured to make and abide by the determination of 1845, with respect to Dover. And if the rule "ne sutor ultra crepidam" be a good one—and a good one it is,—there is still a better, viz., that "there is no rule without exception;" and I claim to be an exception; and to be, though not an engineer by profession, the first Harbour-Engineering Authority in the Country; *and*, I believe I may say, for reasons which I doubt not, will be sufficiently convincing to the reader—if he be anything but a Red-Tape Official—*out of it:* a thing, indeed, that one might be and not have much to boast of after all: since "Observation and Experience have surprisingly failed to produce those results" in the mind of Harbour Engineers, which have lightened up every other circle of Science but theirs. They must be a stolid lot, those Engineer Harbour Makers, although they *are* all, *all* "eminent engineers."

By the bye, it occurs to one, here to ask—Why must Engineers be always "eminent," Eminent Engineers? Is it to express a fact or for the pleasure of re-iterating the letter *E.* that the custom obtains? If the power of alliteration *is* great, must Sense be always required, when Civil Engineers are the persons alluded to, to give way to Sound. Might it not, sometimes, for example, be suitable to substitute the word "execrable" for eminent; or, to use the jingle fittingly to suit the occasion, and at the same time to avoid the libel while holding hard by the profit of the alliteration, get much nearer to the realities of the case: as thus Eminent Engineers of Execrable Engineering! This is indeed an Eminence—an Apex—the Climax —at once the highest Height, and, wonderful paradox, the lowest Depth;—and if a lower can be found or expressed, they deserve the stigma of it.

Eminent Engineering for 150 years!

The result?

"THE BRITISH COAST, THE GRAVE OF THE BRITISH SEAMAN."

The writer of that memorable sentence, on which I have continually rung the changes, and mean again and again to do so, will doubtless quite agree with my alliteration, and with me, in laying the blame at the door of those who have left it to be said, that "the British Coast is the Disgrace of the British Nation:" for "disgrace" implies blame. With whom lies then the Blame? Manifestly with the Admiralty Eminent Engineers; the first Harbour Engineering Authority in the Country. Eminent Engineers of Execrable Engineering!

If by their fruits we are to judge of things and men, to the fruit that is yearly harvested by the Admiralty Engineering, have we a right to look for the character of the Admiralty Engineers.

And what is the Fruit?

Contemplate it in the motto to Paper No. 10.

No. 10.

Facta non Verba.

"The Annual Loss of Property from Casualties on our Coast has been estimated at £1,500,000 sterling. * * * In one year alone, no fewer than 1549 persons perished from these causes alone."—Report of Harb. Ref. Comm., March 1859.

"The British Coast is the Disgrace of the British Nation, and the Grave of the British Seaman."—Quoted by Calver.

It is, however, on grounds, such as have been stated, that the humble individual who writes chooses to exercise his privilege of free speech and of his royal citizenship—*nulli secundus*,—to claim, *under present manifestations*, in the sight and hearing of his Countrymen and the Nation, to be, the First Harbour Authority in the Country, whether with or without Diploma.

"The first harbour authority, in the country or out of it!" "These be brave words, my masters." Indeed are they, but they are such as a man has a right to use, who has "taken off his armour;"—that is to say—whose words 13 years since put on record, have been proved by the event;—2nd, whose judgment in the most recent case, a very important one, has been accepted, *if not acknowledged*, by the compliment of the adoption of his plans by the "official highest marine engineering authority;" and 3rdly, whose predictions of six months standing have received such fulfilment—in what may be considered in one sense, viz., the Cosmopolitan, on account of the many Nations and Interests, which are concerned in its prosperous issue, the most important case of all,—that one hardly knows whether to laugh or to cry at the absurdity of the spectacle presented by the singular occurrence that the realities of the case—stranger than any Fiction that could have been invented—offer, at this moment, to the observation of the Nations.

The justice of the first expression above, is established by the Pamphlet and the Estimate which accompanied it and here precedes it:—

—— of the Second, by the case of Galway Harbour, now before the House and Country:—

—— of the Third, by that of the Danube, upon the importance of the Imperial Parliament's coming to a right resolution concerning which, at the present moment, the facts of the case, as given below, may be held sufficiently to speak for themselves.

But, "the First Harbour Engineering Authority!" H'm. This is rather too much! One *has* heard, that "there is but one step from the sublime to the ridiculous."

It *may*, very possibly, strike some, that while going for the former, the writer of these pages has eminently attained to the latter in the vain-glorious boast he has made. Not at all: nothing like it. I did not *go* for the former: it has come to me. I can not have the latter, it is monopolized. Nevertheless, I know very well how grotesque it must seem for a man to have the assurance to speak of himself as above. But I don't speak for myself; and it is in that, exactly, that the whole case lies. Facts speak, not I. I have merely observed and proclaimed the operations of Nature. *They* bear me out, and more than bear me out in this foolish boasting. So I am not a bit afraid, either of the Grotesque or of the Ridiculous. The facts of the case, of the Coast and the Wrecks, is a proverb, and the accessibility and safety of an Island Harbour is a manifest and self-evident fact.

But if I were afraid of making myself appear contemptible, in my desire to do the good of arousing men to see what the case really is and its true remedy, "is there not a cause," for a man, if need were, to run a risk of the sacrifice of more than personal reputation, in a matter which touches, and may be the means of saving, the lives of thousands, the thousand of human beings yearly lost at sea ; lost by means of the state of our Coasts ? that Coast of which it has been said truly, too truly and even too faintly said, as I have re-quoted above and re-iterate "the British Coast is the disgrace of the British Nation, and the Grave of the British Seaman." This then is a cause in which a man may well sacrifice himself, if so be that any good can be wrought to the victims of *insouisance*, by his devotion ; and I am, therefore, quite willing to run the risk of making myself look ridiculous in the sight of many, if I shall be able to secure the attention of the really influential, to the sad reality that forces itself on the pained vision of the thoughtful, and work their conviction as to the means of abating the evil. I say, that the lives of thousands, and millions of property,—that touches the prosperity of many, many many thousands—are yearly at stake ; and that the greater part of these *are* sacrificed —*that's* the Sacrifice—to the Incapacity and the Presumption of Professional Men :—men, to whom is allowed unlimited scope for mischief of this sort, by the want of Administrative Skill on the part of the Executive, and the Ignorance in things Physical, to which Men in Office unhesitatingly confess. Men, "Men in Office" say, they "see it all, feel it all, but that it must be ; they "can do nothing but what has been done : that if wrecks *are* "numerous so is the Shipping ; and that collisions occur, and "Ships are lost in consequence : that everything is done that "can be done, and that we must be content to grin and bear it."

I say, that everything has *not* been done that can be done. The most simple, manifest, proved means of Harbour Making is disregarded, unattempted, has been denied, mocked at, resisted by those whose DUTY it was to have GRASPED at the faintest idea of good in such a case, in such a hope.

I say, on the other hand, that, for 50 years, with but very rare exceptions, there has been nothing but spending money, in most cases, for worse than nothing, and, but doing to *un*do.

The cause?

MANIFESTLY, because the engineers, when asked (see Blue Books, passim) if they were able to enunciate any general principle of treatment of, or for, harbours, have invariably hedged off the question and said, practically—No. For they have, almost to a man, declared that there is *no* general principle to apply, in that every case had to be treated *sui generis*. And such men are our Engineering Generals! and the result has been exhibited by Admiral Berkeley's Letter, and Tynemouth, and Sunderland, and Dover, and every place of which the quacks have attempted the cure: the cure, on that "old plan" so justly stigmatized by Capt. Elliott, of "fighting Nature;" . . "fighting nature, never very easily done, and has failed in this instance:"—that is, in *every* instance.

At the fully developed height of all this mismanagement, a man comes forward and maintains a General Principle of Treatment, and declares for it that it is of universal application; an Island Breakwater any where, and, *every where*, practicable.

The Authorities, great and small, look at him; they don't ponder on what he says, or how to do it, but stare at *him*. They find he has neither Cocked-Hat on his head, nor a C.E. diploma tacked on to his name; above all this, that he has not the privilege of the entrèe of the magic circle of "the-Admiralty-licensed-to-think;"—nor has he a handle to his name;—neither—and still worse—has he Parliamentary Influence. It is monstrous! An actual Pariah—a Pandy in white!—Hunt him down,—Give him a bad name,—Call him "a busybody in other men's matters," Ask "'Whereunto will this thing grow,' should the intelligent be allowed to speak?" Cry out "'The craft will be in danger' if the Authorized Officials should have to sing-small in admitting an Un-diploma'd Teacher."

Nevertheless, in the progress of the chase-redtape-ical, the taboed malignant comes across the path-official of an Engineer Officer, stranger to him, who happens to be on duty at the place to which he was directed—almost as if in mockery—to turn his steps.

The Officer sees the Plan, appreciates its general applicability,—and is able to pronounce of it, to a High Official, at the time that he is called upon to give evidence before a Harbour Committee in the House, that "it is right in principle, and would answer in 10 places out of 11."

Now, setting aside the want of diploma and all that, here there is " no uncertain sound." A harbour, with two entrances—a plan of treatment, and mode of carying it out, that can be universally applied to the places where it is wanted on that "British Coast," which is now "the Disgrace of the British Nation and the Grave of the British Seaman."

"10 cases out of 11!" nothing of this kind has been seen in England hitherto. 1 out of 10 is as much as anybody can show of successful treatment: and, if this is done under "the highest engineering authority," and all that can be done,—*and it is more than is done*,*—it is manifest that that that promises 10 out of 11 is, must be, the suggestion *of the greatest authority extant.*

Now if it be a joke that the writer has proclaimed of himself, as above, viz., that he has a right to claim to be the First Harbour Engineering Authority in the Country—*the greatest authority extant*—which of course it is, it is a very grim one: grim, that there should be any reasonable ground for saying, even in joke, that a man who is no Engineer is a better Harbour Maker than any in a Maritime Country:—a thing that, be it noted, if true, would by

---

* Harwich is supposed to be this 1 out of 10. Let us come a little near to it.

2390. "Have you visited Harwich lately?" "Yes. I was there last September." 2391. "Has a great amount of refuge been attained by the works which have been carried out there lately?" "Yes, a very material amount since 1838, when I made a survey of it." 2392. "Are all the works finished there?" "Yes, and the Harbour is made available for line-of-battle ships even at half-tide, whereas at the time I was there, there was only 11 ft. at low water, and now it is a magnificent Harbour fit for any thing." What, to allow a man-of-war to take refuge in a storm at low-water? What allowance for the sand of the sea? It is to be remembered also that Harwich is the Admiralty's "one swallow" harbinger of summer.

no means exalt anybody—that country being England; for, its "coasts are a disgrace to the nation"—while the assertion is the shame of those who *profess* marine engineering. I say then, if *this* be a joke, *that*, that that man, whoever he was, wrote is no joke. He wrote it, and he wrote truly. The Wreck-Sheet, published by the Board of Trade, shows that the Coast *is* the Grave of the British Seaman: and as it shows not only that he does lose his life, but *where* he loses it, that points to the *disgrace* of the Admiralty Engineers or Engineer Advisers. These then, *the wrecks at the Harbour's mouths*, are the FACTA of the Admiralty Engineering Authorities; these are their achievements; these Wreck Crosses are the medals with which their "Eminent Engineers" may rejoice to decorate themselves. These are the insignia of their Victory. A joke? It's a joke that relates to Life and Property, and the Prosperity of multitudes. Its converse is Death, Destruction, and Woe!

Now, Gentlemen Sneerers! You, to whom the care of the British Coast is committed,—as Old Dr. Rowland Hill once said to some scoffers that he caught in a pew behind his pulpit, turning round upon them sharp,—"There's fun for you, ladies," so I say to you; "How do you like *my* Sublime and *your* Ridiculous." The greatest good, by the most simple means, and in every place—that's Sublime. Not indeed *my* sublime, but NATURE'S SUBLIME. Look *now* at the Irish Sea,—look at Dover,—look at Sunderland. Have you had enough of the Ridiculous? *Your* RIDICULOUS, *your* DISGRACE, not mine.

I would, therefore, recommend you for the future, while you bear in mind that there are two sides to the picture, not to indulge in the disparagement of the Non-Professional, but to remember that the mass of Wrecks takes place, NOT by collisions at sea, but at the Harbours' Mouths.

"The annual loss of property from casualities on our Coast has been estimated at £1,500,000 sterling, while the average loss of life resulting from them during a period of six years, from 1852 to 1857, has amounted to 780 persons annually. In one year alone (1854), no fewer than 1549 persons perished from these casualties."—*Report of Commissioners, 3rd March*, 1859.

Is that "sublime"? If it be, let me be well-content to be "ridiculous" in pointing out the evil, THE WRONG, and the remedy; while you rejoice to obstruct the action of, and deride, the judgement of all who have not, like yourselves, the uncontrolled privilege of Waste and Squander.

Be it so! Let the Country judge between us.

You are at the Bar of Public Opinion. This is a Case that belongs to the Nation. It is entrusted to your management, and half the wrecks occur at the Harbours' Mouths—at the Haven where the Ship should be—the place the Sailor has reached in the execution of his duty, of the Trust reposed in him to execute. He has done *his* duty. How have you done *yours?* You have brought him to the "Wrecking Pool," &c., &c. *There* is his "Wrecking Pool," produced by your "administrative skill": to the "dangerous decoy,"—as one of your own servants has most justly expressed the fact,—set up as to lure "the British Seaman to his Grave."

Do you dislike this imputation, or desire to escape the responsibility of such charge from this time?

Do then what you ought to do. Change your Plan. Cease to make Pier Harbours. Learn from Nature, while thankful for any guidance that Providence sends you; *so* will be the Wrecks reduced by one-half.

Sometimes people are very much afraid of the Responsibility of Doing.

Is there such a thing as the Responsibility of *Not*-Doing?

I say then,—*Obstruct* what ought to be done, and you are morally guilty of all the Life that is every year lost, at every Port, by reason of Inaccessibility of Harbours.

NOTICE.—The Paper on the Danube will form a separate Publication, and with a Map shewing at one View all the relative Plans of Treatment, will cost 1*s.* 6*d.*